Topics in Group Rings

MONOGRAPHS AND TEXTBOOKS IN PURE AND APPLIED MATHEMATICS

1. *K. Yano,* Integral Formulas in Riemannian Geometry (1970)
2. *S. Kobayashi,* Hyperbolic Manifolds and Holomorphic Mappings (1970)
3. *V. S. Vladimirov,* Equations of Mathematical Physics (A. Jeffrey, editor; A. Littlewood, translator) (1970)
4. *B. N. Pshenichnyi,* Necessary Conditions for an Extremum (L. Neustadt, translation editor; K. Makowski, translator) (1971)
5. *L. Narici, E. Beckenstein, and G. Bachman,* Functional Analysis and Valuation Theory (1971)
6. *D. S. Passman,* Infinite Group Rings (1971)
7. *L. Dornhoff,* Group Representation Theory (in two parts). Part A: Ordinary Representation Theory. Part B: Modular Representation Theory (1971, 1972)
8. *W. Boothby and G. L. Weiss (eds.),* Symmetric Spaces: Short Courses Presented at Washington University (1972)
9. *Y. Matsushima,* Differentiable Manifolds (E. T. Kobayashi, translator) (1972)
10. *L. E. Ward, Jr.,* Topology: An Outline for a First Course (1972) *out of print*
11. *A. Babakhanian,* Cohomological Methods in Group Theory (1972)
12. *R. Gilmer,* Multiplicative Ideal Theory (1972)
13. *J. Yeh,* Stochastic Processes and the Wiener Integral (1973) *out of print*
14. *J. Barros-Neto,* Introduction to the Theory of Distributions (1973) *out of print*
15. *R. Larsen,* Functional Analysis: An Introduction (1973)
16. *K. Yano and S. Ishihara,* Tangent and Cotangent Bundles: Differential Geometry (1973)
17. *C. Procesi,* Rings with Polynomial Identities (1973)
18. *R. Hermann,* Geometry, Physics, and Systems (1973)
19. *N. R. Wallach,* Harmonic Analysis on Homogeneous Spaces (1973)
20. *J. Dieudonné,* Introduction to the Theory of Formal Groups (1973)

21. *I. Vaisman,* Cohomology and Differential Forms (1973)
22. *B.-Y. Chen,* Geometry of Submanifolds (1973)
23. *M. Marcus,* Finite Dimensional Multilinear Algebra (in two parts) (1973, 1975)
24. *R. Larsen,* Banach Algebras: An Introduction (1973)
25. *R. O. Kujala and A. L. Vitter (eds.),* Value Distribution Theory: Part A; Part B. Deficit and Bezout Estimates by Wilhelm Stoll (1973)
26. *K. B. Stolarsky,* Algebraic Numbers and Diophantine Approximation (1974)
27. *A. R. Magid,* The Separable Galois Theory of Commutative Rings (1974)
28. *B. R. McDonald,* Finite Rings with Identity (1974)
29. *J. Satake,* Linear Algebra (S. Koh, T. Akiba, and S. Ihara, translators) (1975)
30. *J. S. Golan,* Localization of Noncommutative Rings (1975)
31. *G. Klambauer,* Mathematical Analysis (1975)
32. *M. K. Agoston,* Algebraic Topology: A First Course (1976)
33. *K. R. Goodearl,* Ring Theory: Nonsingular Rings and Modules (1976)
34. *L. E. Mansfield,* Linear Algebra with Geometric Applications (1976)
35. *N. J. Pullman,* Matrix Theory and its Applications: Selected Topics (1976)
36. *B. R. McDonald,* Geometric Algebra Over Local Rings (1976)
37. *C. W. Groetsch,* Generalized Inverses of Linear Operators: Representation and Approximation (1977)
38. *J. E. Kuczkowski and J. L. Gersting,* Abstract Algebra: A First Look (1977)
39. *C. O. Christenson and W. L. Voxman,* Aspects of Topology (1977)
40. *M. Nagata,* Field Theory (1977)
41. *R. L. Long,* Algebraic Number Theory (1977)
42. *W. F. Pfeffer,* Integrals and Measures (1977)
43. *R. L. Wheeden and A. Zygmund,* Measure and Integral: An Introduction to Real Analysis (1977)
44. *J. H. Curtiss,* Introduction to Functions of a Complex Variable (1978)
45. *K. Hrbacek and T. Jech,* Introduction to Set Theory (1978)
46. *W. S. Massey,* Homology and Cohomology Theory (1978)
47. *M. Marcus,* Introduction to Modern Algebra (1978)
48. *E. C. Young,* Vector and Tensor Analysis (1978)
49. *S. B. Nadler, Jr.,* Hyperspaces of Sets (1978)
50. *S. K. Sehgal,* Topics in Group Rings (1978)

Topics in Group Rings

SUDARSHAN K. SEHGAL

Department of Mathematics
University of Alberta
Edmonton, Alberta, Canada

MARCEL DEKKER, INC.   New York and Basel

Library of Congress Cataloging in Publication Data

Sehgal, Sudarshan K. [Date]
Topics in Group Rings

(Pure and applied mathematics; v. 50)
Bibliography: p.
Includes index.
1. Group rings. I. Title.
QA171.S512 512'.22 78-13671
ISBN 0-8247-6755-1

MARCEL DEKKER, INC.
270 Madison Avenue, New York, New York 10016

Current printing (last digit):
10 9 8 7 6 5 4 3 2 1

PRINTED IN THE UNITED STATES OF AMERICA

PREFACE

The group ring KG of the group G over a commutative unital ring K is an interesting object of study. This is one of the few algebraic structures where one can make explicit computations. Also, there are several easily formulated questions. Historically, group algebras of finite groups appear in the study of representation theory of groups, for example, in the work of Frobenius. But more recently, group rings of infinite groups have been studied in their own right and many interesting results have been obtained by using deep results and techniques from group theory, ring theory and number theory. Great impetus was given to the subject by inclusion of material in books by Lambek and Ribenboim and the two books by Passman. Also, there exist books by Bovdi (Russian) and Zalesskii-Mikhalev (in Russian, translated into English).

The main topics covered in this monograph include the unit group, Lie properties, idempotent problems and the isomorphism problems in KG where K is an arbitrary ring. Most of the material is taken from published literature but there are new results and new proofs here and there. It is hoped that the material will be useful to the advanced graduate student as well as the established mathematician.

The reader is assumed to be familiar with basic group theory including the Sylow theorems, ring theory including the Wedderburn structure theorems and elementary representation theory. Further group theory and commutative algebra results have been developed as required. Also, we have used the Dirichlet Unit Theorem and the Frobenius Density Theory.

Chapter I begins with Zalesskii's theorem, regarding the first coefficient of an idempotent, and its extension by Bass. Results of Formanek, Cliff, Parmenter and the author regarding the absence of idempotents are presented. Also Burns' proof of the theorem regardin the finiteness of the support group of a central idempotent is given.

Chapter II deals with the group of units $\mathcal{U}KG$ of KG. Some basic results regarding torsion units are proved. G. Higman's theor computing the unit group of the commutative integral group ring is given. Integral group rings of finite groups with trivial torsion unit groups and with normal subgroups consisting of trivial units are considered. Also results regarding integral group rings whose units are of finite bounded exponent over the center are proved for use in Chapter VI.

In Chapter III the first isomorphism problem as to whether the isomorphism $RG \simeq RH$ implies $G \simeq H$ is studied. In the next chapter the second isomorphism problem as to whether the isomorphism $R\langle x\rangle \simeq S\langle x\rangle$, $\langle x\rangle$ infinite cyclic, implies $R \simeq S$ is investigated.

In Chapter V the Lie properties, Lie nilpotence, Lie solvabilit Lie n-Engel property, of KG are looked into and the connection with corresponding properties of the unit group $\mathcal{U}KG$ pointed out.

In Chapter VI we begin with the question as to when KG can have nilpotent elements. Then we describe KG whose unit groups are trivial, solvable, nilpotent, FC etc.

The last chapter contains a list of research problems, some of them, no doubt, easy and several very difficult.

Regarding notation, the square brackets refer to the bibliograph at the end. Results within a chapter are referred to by two digits while a three point reference like (I.2.3) refers to (2.3) of Chapte

I wish to express sincere appreciation to Gerald Cliff, Michael Parmenter, James Roseblade and Robert Sandling for reading through parts of the manuscript, making valuable suggestions, and for stimulating discussions; to Donald Passman for encouragement throughout the time I was working on this project; and to my teacher Professor Hans Zassenhaus from whose influence, mathematical and personal, I have benefited over the years.

S.K. SEHGAL

CONTENTS

Topics in Group Rings

CHAPTER I

IDEMPOTENTS

The group ring of a group G over a ring K, with identity, is the ring KG of all formal sums

$$\lambda = \sum_{g \in G} \lambda(g)g, \qquad \lambda(g) \in K,$$

such that

$$\operatorname{supp}(\lambda) = \{g: \lambda(g) \neq 0\},$$

the support of λ, is finite; with the following operational rules

(1) $$\sum_{g \in G} \lambda(g)g = \sum_{g \in G} \mu(g)g \iff \lambda(g) = \mu(g) \quad \text{for all} \quad g \in G,$$

(2) $$\sum_{g \in G} \lambda(g)g + \sum_{g \in G} \mu(g)g = \sum_{g \in G} \bigl(\lambda(g) + \mu(g)\bigr)g,$$

(3) $$\left(\sum_{g \in G} \lambda(g)g\right)\left(\sum_{g \in G} \mu(g)g\right) = \sum_{g \in G} \nu(g)g \quad \text{where}$$

$$\nu(g) = \sum_{h \in G} \lambda(h)\mu(h^{-1}g) = \sum_{xy=g} \lambda(x)\mu(y).$$

Dropping the zero components of the formal sum λ we may write $\lambda = \Sigma_{i=1}^{n} \lambda_i g_i$. Thus $k \to k \cdot e_G$ is an imbedding of K in KG.

After the identification of K with Ke_G we shall assume that K is contained in KG. Clearly, $kg = gk$ for all $k \in K$, $g \in G$. The element $1 \cdot e_G$ acts as identity for KG and we denote it by 1, as usual. The map

$$KG \ni \sum_{g \in G} \lambda(g)g \to \sum_{g \in G} \lambda(g) = c(\lambda) \in K$$

is a homomorphism called the augmentation map of KG. Its kernel,

$$\Delta_K(G) = \left\{ \lambda = \sum_{g \in G} \lambda(g)g \in KG: \sum_{g \in G} \lambda(g) = 0 \right\}$$

is called the augmentation ideal of KG. For a normal subgroup N of G we have the natural homomorphism

$$KG \ni \lambda = \sum_{g \in G} \lambda(g)g \to \sum_{g \in G} \lambda(g)gN \in K(G/N).$$

The kernel of this map,

$$\Delta_K(G,N) = \left\{ \sum_{g \in G} \lambda(g)g \in KG: \sum_{x \in N} \lambda(gx) = 0 \ \text{ for all } \ g \in G \right\}$$

$$= \langle x-1: x \in N \rangle_{KG},$$

is the ideal generated by all $x-1$, $x \in N$. Thus $\Delta_K(G) = \Delta_K(G,G)$. Whenever the coefficient ring K is understood, we shall drop the subscript K and simply write $\Delta(G,N)$ for $\Delta_K(G,N)$. Unless otherwise stated K will always have a 1. We shall denote by $\mathbb{Z}$, $\mathbb{Q}$, $\mathbb{R}$ and $\mathbb{C}$ the rational integers, rational, real and complex numbers respectively.

We now prove a result of Zalesskii and its extension by Bass.

1. IDEMPOTENTS IN QG

Let K be a field of characteristic 0 and G a finite group. Let $e = \Sigma e(g)g$ be an idempotent element of KG. Let us view e in the regular representation of KG, which is given by $x \to R_x$ where R_x is the linear transformation of KG given by

$R_x(y) = xy$ for $x,y \in KG$. Then $R_e = \Sigma e(g)R_g$. The matrix of R_g relative to the basis of KG over K consisting of elements of G, has zeros along the diagonal whenever g is a non-identity element of G. Therefore, taking the matrix trace we have

$$tr(R_e) = \Sigma e(g)tr(R_g) = e(1)|G|.$$

Also, since the characteristic roots of R_e are 0 and 1, $tr(R_e)$ is some non-negative integer r no bigger than $|G|$. Moreover, $r = 0$ (resp. $|G|$) if and only if $e = 0$ (resp. 1).

It is customary to call an idempotent e nontrivial if $e \neq 0,1$. Thus we have proved the following result.

Proposition 1.1. Let K be a field of characteristic 0 and let G be a finite group. If $e = \Sigma e(g)g \in KG$ is a nontrivial idempotent then e(1) is a rational number lying strictly between 0 and 1.

This proposition is also true for arbitrary groups but is considerably more difficult to prove.

Theorem 1.2 (Zalesskii [1]). Let $e = \Sigma e(g)g$ be an idempotent in KG where G is an arbitrary group and K is a field of any characteristic. Then e(1) belongs to the prime subfield of K.

This was conjectured for the characteristic 0 case by Kaplansky who also proved the next result.

Theorem 1.3 (Kaplansky). Let K be a field of characteristic 0 and G any group. If $e = \Sigma e(g)g$ is a nontrivial idempotent of KG then e(1) lies strictly between 0 and 1.

Proof: See Passman ([3], page 93).

In order to prove Theorem 1.2 we need several preliminary results, which we give next. Let us denote for any ring R by [R,R] the additive subgroup of R generated by all Lie products $[x,y] = xy - yx$, $x,y \in R$. The Lie product [x,y] should not be confused with the group commutator $(g,h) = g^{-1}h^{-1}gh$ in a multiplicative group.

<u>Lemma 1.4</u>. Let R be a ring of prime characteristic p, having an identity. Then for $x,y \in R$ and natural numbers n we have

$$(x+y)^{p^n} \equiv x^{p^n} + y^{p^n} \mod [R,R].$$

<u>Proof</u>: Let us first consider the case $n = 1$. Expanding by the distributive law,

$$(x+y)^p = x^p + y^p + \Sigma z_1 z_2 \cdots z_p, \quad z_i \in \{x,y\}, \tag{*}$$

where the sum is over all products $z_1 z_2 \cdots z_p$ of p terms, $z_i \in \{x,y\}$ not all equal to x or y. With each word $z_1 z_2 \cdots z_p$ associate its cyclic permutations

$$z_1 z_2 \cdots z_p, \quad z_2 z_3 \cdots z_p z_1, \quad \cdots, \quad z_p z_1 z_2 \cdots z_{p-1},$$

each differing from the next by a Lie product, for example,

$$z_1 z_2 \cdots z_p - z_2 z_3 \cdots z_p z_1 = [z_1, z_2 z_3 \cdots z_p].$$

It follows that the sum of these cyclic permutations is $pz_1 z_2 \cdots z_p$ modulo $[R,R]$ and thus it belongs to $[R,R]$. Therefore we have from (*)

(i) $(x+y)^p = x^p + y^p + \alpha, \quad \alpha \in [R,R]$.

A consequence of (i) is

(ii) $\gamma \in [R,R] \Rightarrow \gamma^p \in [R,R]$.

This can be seen by observing that

$$\begin{aligned}(xy-yx)^p &\equiv (xy)^p - (yx)^p \mod [R,R] \\ &\equiv [x,(yx)^{p-1}y] \mod [R,R].\end{aligned}$$

The lemma now follows from (i) and (ii) by induction on n. □

Denoting by $\sim$ the conjugacy relation in the group G, write for an element $\alpha = \Sigma\alpha(g)g$ of the group ring RG

$$\tilde{\alpha}(g) = \sum_{h\sim g} \alpha(h).$$

Lemma 1.5. $\alpha \in [RG,RG] \Rightarrow \tilde{\alpha}(x) = 0$ for all $x \in G$.

Proof: From

$$\left[\sum_g \beta(g)g, \sum_h \gamma(h)h\right] = \sum_{g,h} \beta(g)\gamma(h)[g,h]$$

$$= \sum_{g,h} \beta(g)\gamma(h)(gh-hg),$$

it follows that $[RG,RG]$ is the R-linear span of $(gh-hg)$ where $g,h \in G$. Since gh and hg $(= g^{-1}ghg)$ are conjugate we conclude that $\tilde{\alpha}(x) = 0$ for all $x \in G$.

Theorem 1.6. Let R be a commutative ring, with identity, of prime characteristic p. Let $e = \Sigma e(g)g \in RG$ be an idempotent. Then

(i) (Zalesskii) $e(1) = \tilde{e}(1)$ satisfies $e(1)^p = e(1)$;

(ii) (Bass) $\tilde{e}(g^p) = \tilde{e}(g)^p$ for all $g \in G$;

(iii) (Bass) Each $\tilde{e}(g)$ is algebraic over $\mathbb{Z}/(p)$, the integers mod p.

Proof: (i) is a special case of (ii) which we prove first. By (1.4) we have

$$e = e^p = \sum\{e(g)\}^p g^p + \gamma, \quad \gamma \in [RG,RG].$$

This implies, due to (1.5), that

$$\tilde{e}(g) = \sum_{h^p\sim g} e(h)^p = \sum \tilde{e}(h_i)^p \qquad (*)$$

where the last sum is over certain h_i such that $h_i^p \sim g$. For $g \in G$, let us denote by $c\ell(g)$ the conjugacy class of g. Write $S = \{c\ell(g)\colon \tilde{e}(g) \neq 0\}$. Then due to (*), for each $c\ell(g) \in S$, there exists an $h \in G$ with $c\ell(h^p) = c\ell(g)$ and $c\ell(h) \in S$. Hence we can define a map $\psi: S \to S$ by $\mathcal{cl}(h) \xrightarrow{\psi} \mathcal{cl}(h^p)$. This map is then onto and therefore also one to one since S is finite. Thus given $\mathcal{cl}(g) \in S$, there exists a unique $\mathcal{cl}(h) \in S$ with $\mathcal{cl}(h^p) = \mathcal{cl}(g)$. Taking $g = x^p$, (*) gives (ii), namely, $\tilde{e}(g) = \tilde{e}(x^p) = \tilde{e}(x)^p$.

Considering ψ as a permutation of S, $\psi^m = 1$ for some m. Thus

$$\mathcal{cl}(x^{p^m}) = \mathcal{cl}(x)$$

and consequently,

$$\tilde{e}(x) = \tilde{e}(x^{p^m}) = \tilde{e}(x)^{p^m}$$

proving (iii) of the theorem. □

Part (i) of the last theorem proves (1.2) for the characteristic $p > 0$ case. We shall see in (1.13) that this gives the characteristic 0 case by using a deep number theoretic result. But first we need a couple of fundamental theorems of commutative algebra. These results involve finitely generated rings. We shall apply them to $\mathbb{Z}[e(g), g \in G]$, the subring of $\mathbb{C}$ generated by 1 and the support of e.

1.7. Noether's Normalization Theorem

Let $K[x_1,\cdots,x_n] = K(X)$ be a finitely generated integral domain over a field K. Then there exist elements $y_1,\cdots,y_r \in K[X]$, $r \geq 0$, algebraically independent over K, such that $K[X]$ is integral over $K[y_1,\cdots,y_r] = K[Y]$.

Proof: We use induction on n. If $x_1,\cdots,x_n$ are algebraically independent over K there is nothing to prove. We may suppose, therefore, that there is a nontrivial relation

$$\sum a_{j_1 j_2 \cdots j_n} x_1^{j_1} x_2^{j_2} \cdots x_n^{j_n} = 0 \qquad (*)$$

with each $a_{j_1 j_2 \cdots j_n}$ a nonzero element of K and each $j_i \geq 0$.

Let us make the substitution

$$x_i = y_i + x_1^{m_i} \qquad i = 2,\cdots,n$$

in (*), where m_i are natural numbers to be suitably chosen. We get

$$\sum a_{j_1 j_2 \cdots j_n} x_1^{j_1}(y_2+x_1^{m_2})^{j_2}\cdots(y_n+x_1^{m_n})^{j_n} = 0.$$

That is,

$$\sum a_{j_1 j_2 \cdots j_n} x_1^{(j_1+m_2 j_2+\cdots+m_n j_n)} + f(x_1,y_2,\cdots,y_n) = 0 \qquad (**)$$

where $f(x_1,y_2,\cdots,y_n)$ is a polynomial in $x_1,y_2,\cdots,y_n$ over K containing no pure powers of x_1. Notice that the highest power of x_1 in (**) is pure. We wish to choose m_i such that all pure powers of x_1 appearing in (**) are distinct. We first choose a natural number d which is not a root of any equation of the form

$$\sum_{i=o}^{n} c_i z^i = 0$$

where each c_i is a difference of two j_i's appearing in (*). This we can do as there are only a finite number of such equations and some $c_i \neq 0$ in each equation. Now, setting $m_i = d^i$, $i \geq 2$, we find that either x_1 is a root of unity or

$$x_1^{j_1+m_2 j_2+\cdots+m_n j_n}$$

are distinct. As a result (**) implies that x_1 is integral over $K[y_2,\cdots,y_n]$. Hence, $K[X] = K[x_1,y_2,\cdots,y_n]$ is integral over $K[y_2,\cdots,y_n]$. Now by induction $K[y_2,\cdots,y_n]$ contains elements $z_1,z_2,\cdots,z_r$, $r \geq 0$, algebraically independent over K such that $K[y_2,\cdots,y_n]$ is integral over $K[z_1,\cdots,z_r]$. The proof is now complete since an integral extension of an integral extension is integral.

1.8. Extension Theorem

Let $A \subseteq B$ be integral domains. Suppose that B is integral over A. Then any homomorphism $\phi: A \to K$, of A into an algebraically closed field K can be extended to B.

Proof: Let S be the set of all pairs (R,σ) where R is a ring with $A \subseteq R \subseteq B$ and $\sigma: R \to K$ is an extension of ϕ. For $(R_1,\sigma_1),(R_2,\sigma_2) \in S$, we say $(R_1,\sigma_1) \leq (R_2,\sigma_2)$ if and only if $R_1 \subseteq R_2$ and σ_2 is an extension of σ_1. S is non-empty and is inductively ordered. By Zorn's lemma, S has a maximal element (k,λ). We claim that $k = B$. If not, there exists an element $\alpha \in B$, $\alpha \notin k$. It suffices to extend λ to $k[\alpha]$ thereby contradicting the maximality of (k,λ). Let us first write, for convenience, $T = \{t \in k: \lambda(t) \neq 0\}$ and extend λ to the subring k_T, of the field of quotients of B, given by

$$k_T = \{a/t: a \in k,\ t \in k,\ \lambda(t) \neq 0\}.$$

This can be done by setting $\mu(a/t) = \lambda(a)/\lambda(t)$. It is easy to check that μ is a well defined extension of λ. Also, $\mu(k_T)$ is a subfield F of K. This is because

$$\mu(a/t) \neq 0 \Rightarrow \mu(a)/\mu(t) \neq 0 \Rightarrow \mu(a) \neq 0 \Rightarrow \mu(t)/\mu(a) \in \mu(k_T)$$

and $\mu(a/t)$ has an inverse. Clearly μ is an extension of λ.

Now we intend to extend μ to $k_T[\alpha]$. We need to choose $\mu(\alpha)$ suitably. Let us denote by $\Sigma\bar{a}_i x^i = \Sigma\mu(a_i)x^i = \bar{f}(x) \in F[x]$ the image under μ of any polynomial $f(x) = \Sigma a_i x^i \in k_T[x]$. Write

$$I = \{\bar{f}(x) \in F[x]: f(x) \in k_T[x] \text{ and } f(\alpha) = 0\}.$$

I is an ideal of $F[x]$ and contains a nonzero polynomial since α is integral over k_T. So I is a principal ideal generated by, say, $m(x)$. Choose a root $\xi \in K$ of $m(x)$ and set

$$\tau\left(\sum a_i \alpha^i\right) = \sum \mu(a_i)\xi^i$$

for $a_i \in k_T$. Then τ is a well defined homomorphism:

$$\sum a_i\alpha^i = \sum b_i\alpha^i \Rightarrow \sum(a_i-b_i)\alpha^i = 0 \Rightarrow \sum\overline{(a_i-b_i)}x^i \in I$$

$$\Rightarrow \text{m(x) divides } \sum\overline{(a_i-b_i)}x^i \Rightarrow \sum\overline{(a_i-b_i)}(\xi^i) = 0 \Rightarrow \sum\overline{a_i}\xi^i = \sum\overline{b_i}\xi^i.$$

Clearly now τ, restricted to $k[\alpha]$, is the map contradicting the maximality of (k,λ) and proving the theorem. □

Now we apply Theorems 1.7 and 1.8 to obtain a well known result which we use in this chapter as well as later.

Theorem 1.9. Let $R = \mathbb{Z}[\alpha_1,\alpha_2,\cdots,\alpha_n]$ be a finitely generated integral domain and let x be a non-zero element of R. Then there exists an infinite number of primes p such that for each p, there is a field F_p of characteristic p and a homomorphism $\phi_p\colon R \to F_p$ satisfying $\phi_p(x) \neq 0$. Moreover, if x is transcendental over the field of rational numbers Q, the F_p can be chosen so that $\phi_p(x)$ is transcendental over $\mathbb{Z}/(p)$, the field of p elements.

Proof: (a) Suppose that x is transcendental over Q. Taking $K = Q(x)$ and applying Noether's normalization theorem to $K[\alpha_1,\cdots,\alpha_n]$ we conclude that $K[\alpha_1,\cdots,\alpha_n]$ contains elements $(y_1,y_2,\cdots,y_r) = Y$, algebraically independent over K, such that $K[\alpha_1,\cdots,\alpha_n]$ is integral over $K[Y]$. In other words, we have the containments

$$K = Q(x) \to K[Y] \to K[Y][\alpha_1,\cdots,\alpha_n],$$

where the last extension is integral.

Pick a monic polynomial f_i satisfied by α_i over $K[Y]$. Each coefficient of f_i is a polynomial in Y over K i.e. a polynomial in Y whose coefficients are quotients of elements of $\mathbb{Z}[x]$. Let S be the finite set of all primes of $\mathbb{Z}[x]$ appearing as factors of numerators or denominators of the nonzero coefficients of any of the coefficient polynomials of f_i,

$1 \le i \le n$. Then α_i is integral over $\mathbb{Z}[x]_S[Y]$ where $\mathbb{Z}[x]_S$ is the ring of quotients $\left\{\frac{a(x)}{b(x)} : a(x) \in \mathbb{Z}[x],\ b(x) \text{ a product of primes in } S\right\}$. We have

$$\mathbb{Z}[x]_S \to \mathbb{Z}[x]_S[Y] \to \mathbb{Z}[x]_S[Y][\alpha_1, \cdots, \alpha_n]$$

where the second extension is integral. There exists an infinite number of rational primes p which do not divide any coefficient of any polynomial in S. For any such p let us extend the natural homomorphism $\mathbb{Z} \to \mathbb{Z}/(p)$ to

$$\phi_p : \mathbb{Z}[x] \to \mathbb{Z}/(p)\ (t)$$

where t is a transcendental over $\mathbb{Z}/(p)$ by setting $\phi_p(x) = t$. This can be further extended to

$$\phi_p : \mathbb{Z}[x]_S \to \mathbb{Z}/(p)\ (t)$$

and by mapping $y_i \to 0$ we have a homomorphism

$$\phi_p : \mathbb{Z}[x]_S[Y] \to \mathbb{Z}/(p)\ (t)$$

which by Theorem 1.8 can be extended to a homomorphism of $\mathbb{Z}[x]_S[Y][\alpha_1, \cdots, \alpha_n]$ to the algebraic closure of $\mathbb{Z}/(p)\ (t)$. This is the desired homomorphism.

(b) Suppose that x is algebraic over Q. Applying Noether's normalization theorem to $Q[\alpha_1, \cdots, \alpha_n]$ we conclude that $Q[\alpha_1, \cdots, \alpha_n]$ contains elements $(y_1, \cdots, y_r) = Y$, algebraically independent over Q such that $Q[\alpha_1, \cdots, \alpha_n]$ is integral over $Q[Y]$. We have the inclusions

$$Q \to Q[Y] \to Q[\alpha_1, \cdots, \alpha_n].$$

Let

$$f_0(x) = x^m + \frac{a_1}{b_1} x^{m-1} + \cdots + \frac{a_m}{b_m} = 0,\ a_i b_i \in \mathbb{Z},\ a_m \ne 0 \qquad (*)$$

be the minimal polynomial of x over Q. Let f_i be a monic polynomial satisfied by α_i, $i \ge 1$, over $Q[Y]$. Let S be the finite set of rational primes which divide the numerator or denominator of any coefficient of f_i for any $i \geq 0$. Suppose $\mathbb{Z}_S$ is the ring of quotients,

$$\mathbb{Z}_S = \{a/b\colon a \in \mathbb{Z},\ \ b \text{ a product of primes in } S\}.$$

Then we have

$$\mathbb{Z}_S \to \mathbb{Z}_S[Y] \to \mathbb{Z}_S[\alpha_1,\cdots,\alpha_n]$$

where the last extension is integral. Let p be any rational prime not in S. Extend the natural projection $\mathbb{Z} \to \mathbb{Z}/(p)$ to a homomorphism

$$\phi_p\colon \mathbb{Z}_S[Y] \to \mathbb{Z}/(p)$$

by mapping each $y_i \to 0$. By Theorem 1.8, ϕ_p can be extended to

$$\phi_p\colon \mathbb{Z}_S[\alpha_1,\cdots,\alpha_n] \to K_p ,$$

where K_p is the algebraic closure of $\mathbb{Z}/(p)$. It follows from (*) that $\phi_p(x) \neq 0$. Since $R \subseteq \mathbb{Z}_S[\alpha_1,\cdots,\alpha_n]$ and S is finite, the proof is complete. □

Theorem 1.10 (Bass). Let $e = \Sigma e(g)g = e^2 \in kG$, where k is a field of characteristic 0. Then

(i) Each $\tilde{e}(g)$ is algebraic over Q;

(ii) There exists a homomorphism $\lambda\colon \mathbb{Z}[e(g), g \in G] \to F$, where F is a field of characteristic 0 such that $\lambda(\tilde{e}(g)) = \tilde{e}(g)$ for all $g \in G$ and $\lambda(e(g))$ is algebraic over Q.

Proof: (i) Suppose that $\tilde{e}(g_0) = x$ is transcendental over Q. Then we know by Theorem 1.9 that there exists a homomorphism

$$\phi_p\colon \mathbb{Z}[e(g), g \in G] \to F_p$$

where F_p is a field of characteristic p such that $\phi_p(x)$ is transcendental over $\mathbb{Z}/(p)$. The element $f = \Sigma\phi_p(e(g))g$ is an idempotent of F_pG. Moreover,

$$\tilde{f}(g_0) = \sum_{h\sim g_0} f(h) = \sum_{h\sim g_0} \phi_p\big(e(h)\big) = \phi_p\left(\sum_{h\sim g_0} e(h)\right) = \phi_p\big(\tilde{e}(g_0)\big) = \phi_p(x)$$

is a transcendental over $\mathbb{Z}/(p)$ which contradicts Theorem 1.6. Hence $\tilde{e}(g_0)$ is algebraic over Q.

(ii) Let $K = Q\big(\tilde{e}(g), g \in G\big)$. Applying Noether's normalization theorem to $R = K[e(g), g \in G]$ we conclude that R contains elements $(y_1, \cdots, y_r) = Y$, algebraically independent over K such that R is integral over K[Y]. We have the inclusions

$$K \to K[Y] \to K[e(g), g \in G].$$

Let λ be the homomorphism which is identity on K and zero on Y. By Theorem 1.8, λ can be extended to

$$\lambda\colon K[e(g), g \in G] \to \hat{K},$$

where $\hat{K}$ is the algebraic closure of K. Clearly $\lambda\big(\tilde{e}(g)\big) = \tilde{e}(g)$ and $\lambda\big(e(g)\big)$ is algebraic over K. Since K is algebraic over Q by the first part of this theorem, $\lambda\big(e(g)\big)$ is algebraic over Q. □

1.11. Remark. One might be tempted to think that every coefficient of an idempotent of a group ring over a field of characteristic 0 is algebraic over Q. However, this is not true in general. Let G be a finite non-commutative group. Let A be the field of all algebraic numbers and $\mathbb{C}$ the field of complex numbers. Then by the Wedderburn structure theorem AG is a finite direct sum of $D^{(i)}_{n_i}$ (the ring of all $n_i \times n_i$ matrices over the division ring $D^{(i)}$), $1 \le i \le s$. But since $D^{(i)}$ are finite dimensional algebras over A and A is algebraically closed, we have that each $D^{(i)} = A$. Thus

$$AG = A_{n_1} \oplus \cdots \oplus A_{n_s}$$

and since AG is not commutative at least one n_i, say n_1, is greater than one. Also,

$$\mathbb{C}G = \mathbb{C}_{n_1} \oplus \cdots \oplus \mathbb{C}_{n_s} .$$

Now, pick t a transcendental in $\mathbb{C}$ and consider

$$e = \begin{pmatrix} 1 & t & 0 & \cdots & 0 \\ 0 & 0 & \cdots & & 0 \\ \vdots & & & & \\ \vdots & & & & \\ 0 & 0 & \cdots & & 0 \end{pmatrix} = e^2 \in \mathbb{C}_{n_1} .$$

The element $(e,0,\ldots,0)$ is an idempotent of $\mathbb{C}G$ whose coefficients are not all algebraic because if they were then $(e,0,\ldots,0) \in AG = A_{n_1} \oplus \cdots \oplus A_{n_s}$ which is not true.

In order to prove Theorem 1.2 we need to use the classical Frobenius density theorem. We recall a few basic facts in order to state this theorem; further details can be found in Janusz [1]. Let K/Q be a finite normal extension with Galois group $G = G(K,Q)$. For a prime $\mathfrak{p}$ of K lying over the rational prime p let us denote by $O_{\mathfrak{p}}$ and $M_{\mathfrak{p}}$ the corresponding valuation ring and the maximal ideal respectively. Denote by $\overline{K}$ the residue class field $O_{\mathfrak{p}}/M_{\mathfrak{p}}$ which has $p^{f_{\mathfrak{p}}} = p^f$ elements. The group G permutes the primes $\mathfrak{p}$ of K which lie over a given rational prime p. The stabilizer of $\mathfrak{p}$ in $G = \{\sigma \in G\colon \mathfrak{p}^{\sigma} = \mathfrak{p}\} = G_{\mathfrak{p}}$ acts on $\overline{K}$. In fact, the implied map

$$\psi_{\mathfrak{p}}\colon G_{\mathfrak{p}} \to G(\overline{K}, \mathbb{Z}/(p))$$

is onto and its kernel $T_{\mathfrak{p}}$ is the inertia group. The order of $T_{\mathfrak{p}}$ is the ramification index $e(\mathfrak{p}/p) = e_{\mathfrak{p}}$. So if $\mathfrak{p}$ is unramified $\psi_{\mathfrak{p}}$ is 1-1 and onto.

Now $\overline{K} = O_{\mathfrak{p}}/M_{\mathfrak{p}}$ is a finite field having $q = p^{f_{\mathfrak{p}}}$ elements and $G(\overline{K}, \mathbb{Z}/(p))$ has a canonical generator, namely, $x \to x^p$. If $\mathfrak{p}$ is unramified, the inverse image under ψ_p of this canonical generator is called the Frobenius automorphism, $\mathrm{Frob}(\mathfrak{p})$, belonging to $\mathfrak{p}$. In other words, $\sigma = \mathrm{Frob}(\mathfrak{p})$ is characterized by

$$\sigma x \equiv x^p \pmod{M_{\mathfrak{p}}} \quad \text{for all} \quad x \in O_{\mathfrak{p}}.$$

1.12. Frobenius Density Theorem

For any $\sigma \in \mathcal{G}(K,Q)$, there is an infinite number of rational primes p such that there exist primes p dividing p with $\text{Frob}(p) = \sigma^m$ and $\langle\sigma^m\rangle = \langle\sigma\rangle$.

Proof: See Janusz [1], page 134.

In order to complete the proof of Theorem 1.2 we have to show that if K is a field of characteristic 0 then $e = e^2 \in KG \Rightarrow e(1) \in Q$. It is convènient to prove this statement together with another result.

Theorem 1.13. Let K be a field of characteristic 0 and $e = \Sigma e(g)g = e^2 \in KG$. Then

(i) (Zalesskii) $e(1) \in Q$;

(ii) (Bass) $E = Q(\tilde{e}(g), g \in G)$ is a finite normal extension of Q with Abelian Galois group.

Proof: Choose a homomorphism $\lambda: \mathbb{Z}[e(g): g \in G] \to F$ as in Theorem 1.10, part (ii) and let $f = \Sigma\lambda(e(g))g = f^2$. Since $\tilde{f}(g) = \lambda(\tilde{e}(g)) = \tilde{e}(g)$, we may replace e by f and thus assume that all e(g) are algebraic over Q. Let N be the normal closure of $k = Q(e(g): g \in G)$ and $\mathcal{G}$ the Galois group of N/Q. It is easy to see that all but a finite number of primes p of N satisfy

(a) $e_g \in O_p$ for all $g \in G$,

(b) For all $g,h \in G$, $\sigma \in \mathcal{G}$,

$$\sigma\tilde{e}(g) - \tilde{e}(h) \neq 0 \Rightarrow \sigma\tilde{e}(g) - \tilde{e}(h) \notin M_p.$$

We claim

(1.14) For any $\sigma \in \mathcal{G}$ there exists an $n = n(\sigma)$ such that $\sigma\tilde{e}(g) = \tilde{e}(g^n)$ for all $g \in G$.

Given $\sigma \in \mathcal{G}$, by the Frobenius density theorem, there exists an m such that $\langle\sigma^m\rangle = \langle\sigma\rangle$ and $\text{Frob}(p) = \sigma^m$ for an infinity of primes p of N. We pick a fixed prime p satisfying (a) and (b) with

$$\langle\sigma^m\rangle = \langle\sigma\rangle, \quad \mathrm{Frob}(p) = \sigma^m.$$

Then $e \in O_pG$ and $\bar{e} \in F_pG$ where $F_p = O_p/M_p$ is a field of characteristic p, $p|p$. Applying (1.6)(ii) we conclude that for all $g \in G$,

$$\tilde{e}(g)^p \equiv \tilde{e}(g^p) \pmod{M_p}.$$

Thus

$$\sigma^m\, \tilde{e}(g) \equiv \tilde{e}(g^p) \pmod{M_p}$$

which implies by (b) that $\sigma^m\tilde{e}(g) = \tilde{e}(g^p)$. If $\sigma^{m\ell} = \sigma$, we apply σ^m ℓ times to get

$$\sigma\, \tilde{e}(g) = \tilde{e}(g^{p^\ell})$$

for all $g \in G$. This proves (1.14).

In particular, if we take $g = 1$ we deduce that $\sigma e(1) = e(1)$ for all $\sigma \in G$. Hence $e(1) \in Q$, proving (i).

It is clear from (1.14) that $\sigma E \subseteq E$ and thus E is normal. Moreover, for $\sigma,\tau \in G$ there exist n, n' such that

$$(\sigma\tau)\tilde{e}(g) = \sigma\, \tilde{e}(g^{n'}) = \tilde{e}(g^{nn'}) = (\tau\sigma)\tilde{e}(g).$$

We have proved that the Galois group of E/Q is Abelian. □

It has been proved in the last theorem that $E = Q(\tilde{e}(g), g \in G)$ is an Abelian extension and thus by the well known theorem of Kronecker and Weber E is contained in a cyclotomic field. Bass [2] has made the following conjecture and shown that it is true for polycyclic by finite groups.

1.15. Conjecture (Bass). $E = Q(\tilde{e}(g), g \in G)$ is contained in $Q(\xi_\ell)$ where ξ_ℓ is a primitive ℓ-th root of unity and ℓ is the least common multiple of all finite orders $o(g)$ of the group elements $g \in G$, for which $\tilde{e}(g) \neq 0$.

2. NOETHERIAN GROUPS

Definition 2.1. A group is said to be Noetherian if it satisfies the ascending chain condition on subgroups i.e. any sequence

$$G_0 \subseteq G_1 \subseteq G_2 \subseteq \cdots \subseteq G_n \subseteq \cdots$$

of subgroups of G becomes stationary after a finite number of steps.

The only known examples of Noetherian groups are polycyclic-by-finite groups i.e. groups G having a normal subgroup N such that G/N is finite and N has a subnormal series

$$N = N_0 \rhd N_1 \rhd N_2 \rhd \cdots \rhd N_s = \{1\}$$

with each factor N_i/N_{i+1} cyclic.

Proposition 2.2. A polycyclic-by-finite group G is Noetherian.

Proof: We know that G has a subnormal series

$$G = G_0 \rhd G_1 \rhd G_2 \rhd \cdots \rhd G_s = \{1\}$$

with each G_i/G_{i+1} finite or cyclic. We may use induction on s. Therefore $G_1 = N$ is Noetherian and clearly G/G_1 is Noetherian. Suppose that

$$H_1 \subseteq H_2 \subseteq H_3 \subseteq \cdots \subseteq H_n \subseteq \cdots$$

is a series of subgroups of G. Then since

$$N \cap H_1 \subseteq N \cap H_2 \subseteq \cdots \subseteq N \cap H_n \subseteq \cdots$$

and

$$H_1N/N \subseteq H_2N/N \subseteq \cdots \subseteq H_nN/N \subseteq \cdots$$

are series of subgroups of N and G/N we have that there exists an i with

$$N \cap H_i = N \cap H_j, \quad H_iN/N = H_jN/N$$

for all $j \geq i$. Now for $j \geq i$,

$$H_j = H_j \cap (H_iN) = H_i(H_j \cap N) = H_i$$

and thus G is Noetherian.

2.3. Hirsch number

Let G be a group having a subnormal series

$$G = G_0 \rhd G_1 \rhd \cdots \rhd G_k = \{1\} \tag{2.4}$$

with each factor G_i/G_{i+1} finite or cyclic. Then the number of infinite cyclic factors in (2.4) is an invariant of G i.e. any other series

$$G = H_0 \rhd H_1 \rhd \cdots \rhd H_\ell = \{1\} \tag{2.5}$$

with H_i/H_{i+1} finite or cyclic has the same number of infinite cyclic factors as (2.4). This will follow if we can show that any refinement of a series like this has the same number of infinite cyclic factors because (2.4) and (2.5) do have isomorphic refinements by the Jordan-Hölder-Schreier-Zassenhaus Theorem (see M. Hall [1]). Now let

$$G = G_0 \rhd \cdots \rhd G_i = N_{i0} \rhd N_{i1} \rhd \cdots \rhd N_{it} = G_{i+1} \rhd \cdots \tag{2.6}$$

be a refinement of (2.4). If G_i/G_{i+1} is finite, each $N_{i,\ell}/N_{i,\ell+1}$ is, of course, finite and if G_i/G_{i+1} is infinite cyclic then $N_{i,t-1}/N_{i,t}$ is the only factor between G_i and G_{i+1} which is infinite cyclic. Consequently (2.4) and (2.6) have the same number of infinite cyclic factors. This proves our assertion. □

The number of infinite cyclic factors in (2.4) is called the Hirsch number of G and is denoted by $h(G)$.

A Noetherian group G satisfies the following property.

(2.7) If an element of infinite order $g \in G$ is conjugate to g^n then $n = 1$ or -1.

Proof: Suppose $xgx^{-1} = g^n$, $x \in G$. Let $h_i = x^{-i}gx^i$ and $H_i = \langle h_i \rangle$. Then

$$h_{i+1}^n = (x^{-i-1}gx^{i+1})^n = x^{-i-1}g^n x^{i+1} = x^{-i}gx^i = h_i$$

and we have an ascending chain

$$H_0 \subseteq H_1 \subseteq \cdots \subseteq H_n \subseteq \cdots$$

of subgroups of G. Therefore for some i, $H_{i+1} = H_i$, i.e. $\langle h_i \rangle = \langle h_{i+1} \rangle$ and $h_{i+1}^n = h_i$. Hence $n = 1$ or -1.

Theorem 2.8 (Formanek [2]). Let $e = \Sigma e(g)g$ be an idempotent of KG where K is any field and G is a group satisfying (2.7). Then

(i) $\tilde{e}(x) = 0$ for all $x \in G$ of infinite order,

(ii) $e(1) = 0$ or 1, if G is torsion free.

Proof: (i) First let us suppose that K has characteristic $p > 0$. Write $x \sim y$ to mean that two elements x, y of G are conjugate. Given $g,h \in G$, of infinite order there is at most one n such that $h^{p^n} \sim g$ because

$$h^{p^n} \sim g,\quad h^{p^{n+i}} \sim g \Rightarrow h^{p^{n+i}} = (h^{p^n})^{p^i} \sim h^{p^n} \Rightarrow i = 0, \quad \text{by (2.7)}$$

Now let an element x, of infinite order, of G be given. Choose n large enough so that no $g^{p^n} \sim x$ for $g \in \mathrm{Supp}(e)$. We have by (1.4),

$$e = e^{p^n} = \sum e(g)^{p^n} g^{p^n} + \beta, \quad \beta \in [KG,KG].$$

This implies (by the choice of n and (1.5)) that $\tilde{e}(x) = 0$. The same now holds if the characteristic of K is 0, in view of (1.9).

(ii) Since e is an idempotent, $(\Sigma e(g)) = (\Sigma e(g))^2 = 0$ or 1. Also,

$$\sum_g e(g) = e(1) + \sum \tilde{e}(x)$$

where the sum is over certain number of elements x of the torsion free group G. Thus $e(1) = 0$ or 1.

Corollary 2.9 (Formanek [2]). If G is a torsion free Noetherian group and K is a field of characteristic 0 then KG has no nontrivial idempotents.

Proof: Theorem 2.8, part (ii) and Theorem 1.3.

The last corollary gives a partial answer to the following well known

Conjecture 2.10. The group algebra KG of a torsion free group over a field has no nontrivial idempotents.

Let K be a field of characteristic 0 and suppose that $e = \Sigma e(g)g$ is a nontrivial idempotent. Then we know by Zalesskii's Theorem (1.2) that $e(1)$ is a rational number r/s, $r, s \in \mathbb{Z}$, $(r,s) = 1$. We make the

Conjecture 2.11. For every prime p dividing s, G has an element of order p.

Two remarks are in order. Firstly, Conjecture 2.10 for the characteristic 0 case follows from Conjecture 2.11 in view of Kaplansky's Theorem (1.3), since a torsion free group has no elements of order p. Secondly, it is not true that if p^α divides s then G has an element of order p^α as can be seen by the following example. Let G be an elementary abelian group of order p^n, then

$$e = \frac{1}{p^n} \sum_{g \in G} g$$

is an idempotent of QG with $e(1) = 1/p^n$ but G has no element of order p^n if $n > 1$.

We offer a proof of Conjecture 2.11 for polycyclic-by-finite groups G. We need a few lemmas.

Lemma 2.12. An infinite polycyclic-by-finite group G contains an Abelian torsion-free normal subgroup $N \neq \{1\}$.

Proof: We have a subgroup A of G such that G/A is finite and A is polycyclic and hence solvable. Let us denote by A_i the i-th term in the derived series of A i.e. A_i is the derived group (A_{i-1}, A_{i-1}) of A_{i-1}. We have the sequence

$$G \supset A = A_1 \supset A_2 \supset \cdots \supset A_n \supset A_{n+1} = \{1\}$$

where $A_n \neq \{1\}$. We claim that

(2.13) G has an infinite Abelian normal subgroup N_1.

Either A_n is infinite, in which case, take $N_1 = A_n$; or A_n is finite, whence by induction on n G/A_n has an infinite Abelian normal subgroup B/A_n. Let t be the order of the automorphism group of A_n. Then g^t centralizes A_n for every $g \in G$. Thus for $x, y \in B$, $x^{|A_n|t}$ commutes with y. Therefore,

$$B^{|A_n|t} = \langle b^{|A_n|t}, b \in B \rangle = N_1$$

is an infinite normal Abelian subgroup of G. Writing $N_1 = F \times T$ where T is finite and F is torsion free, $N_1^{|T|}$ is the desired subgroup N.

Lemma 2.14 (Hattori [1]). Suppose that $e = \Sigma e(g)g$ is an idempotent in the group algebra KG of a finite group over a field of characteristic 0. Let χ be the character of G afforded by KGe. Then for $g \in G$, we have

$$|C_G(g)|\tilde{e}(g) = \chi(g^{-1})$$

where $C_G(g)$ denotes the centralizer of g in G.

Proof: For any $\alpha \in KG$ and $h \in G$, let $T_\alpha(h)\colon KG \to KG$ be the K-linear map which sends y to $h y \alpha$. Then $T_e(h)$ acts on KGe as left multiplication by h and annihilates $KG(1-e)$. Since $KG = KGe \oplus KG(1-e)$, choosing a suitable basis of KG, we see that the trace of the linear transformation $T_e(h)$ is equal to $\chi(h)$.

Now $T_e(h) = \Sigma_j e(g)T_g(h)$ and $T_g(h)$ sends x to $h x g$ for any $x \in G$. Therefore, $T_g(h)$ permutes the elements of G and its trace is the number of elements $x \in G$ with $x = h x g$. But $x = h x g$ if and only if $x^{-1}h^{-1}x = g$. So the trace of $T_g(h)$ is 0 if and only if g is not conjugate to h^{-1} and is $|C_g(h)|$ otherwise. Hence

$$\chi(h) = \text{trace } T_e(h) = \sum_{g\sim h^{-1}} e(g)|C_G(h)|$$

$$= \tilde{e}(h^{-1})|C_G(h)|.$$

Theorem 2.15 (Cliff and Sehgal [1]). Let KG be the group ring of a polycyclic-by-finite group G over a field K of characteristic 0. Let $e = \Sigma e(g)g$ be a nontrivial idempotent. Write the rational number, $e(1) = \frac{r}{s}$ with $(r,s) = 1$. If a prime p divides s then there exists a non-identity element $g \in G$ of p-power order with $\tilde{e}(g) \neq 0$.

Proof: Suppose that G/A is finite and A is polycyclic. We use induction on the Hirsch number $h(G)$. We claim that if p is a prime divisor of s then there is an element $g \in G$, of p-power order with $\tilde{e}(g) \neq 0$, $g \neq 1$. Suppose that $h(G) \geq 1$; then by Lemma 2.12, G has a torsion free normal subgroup $N \neq \{1\}$ and therefore G/N has smaller Hirsch number. Let $\bar{e}$ be the image of e under the natural map $KG \to K(G/N)$. Then due to

Theorem 2.8, $\overline{e}(1) = e(1) = \frac{r}{s}$ because $\overline{e}(1) = \Sigma_{x \in N} e(x) = e(1) + \Sigma\tilde{e}(y)$, where the sum is over certain $y \in N$. Therefore, by induction, there is a $\overline{g} \in G/N$ of p-power order such that

$$0 \neq \tilde{\overline{e}}(\overline{g}) = \sum\tilde{e}(h),$$

a sum over certain $h \in G$ such that $\overline{h}$ is conjugate to $\overline{g}$. Since again due to Theorem 2.8, $\tilde{e}(h) = 0$ for elements h of infinite order, we have that $\tilde{e}(h_0) \neq 0$ for some h_0 of finite order. This h_0 clearly has p-power order. Thus it remains to prove

Lemma 2.16. If G is finite and p is a prime divisor of s then there exists an element $g \in G$ of p-power order with $\tilde{e}(g) \neq 0$, $g \neq 1$.

Proof: Suppose on the contrary that $\tilde{e}(g) = 0$ for all p-elements $g \neq 1$. Let P be a Sylow p-subgroup of G and let χ_P be the restriction to P of χ, the character of G afforded by KGe. Then by Lemma 2.14,

$$\chi_P(g) = 0 \quad \text{for all} \quad 1 \neq g \in P.$$

Therefore,

$$\frac{\chi(1)}{\zeta(1)} \zeta(g) = \chi_P(g) \quad \text{for all} \quad g \in P,$$

where ζ is the character of the regular representation of P. Since the 1-representation occurs once and only once as a component of the regular representation it follows that $\chi(1)/\zeta(1)$ is an integer. Thus $\chi(1)$ is a multiple of $|P|$. But $\chi(1) = |G|\frac{r}{s}$, so p cannot divide s, proving the lemma and hence the theorem.

We now investigate as to when group rings RG, where R is commutative ring with identity, have idempotents $\neq 0,1$. If G has an element g of order $n > 1$ and $n \in \mathcal{U}R$, the unit group of R, then

$$e = \frac{1}{n}\left(1 + g + \cdots + g^{n-1}\right)$$

is an idempotent of RG. So for RG to be without nontrivial idempotents, it is necessary that if g has an element of finite order, $o(g) > 1$ then $o(g) \notin \mathcal{U}R$. If R has zero characteristic this condition is equivalent to

$$(2.17) \qquad \{o(g): g \in G\} \cap \mathcal{U}R = \{1\}.$$

Moreover, one might wish to know that if (2.17) is satisfied, whether or not

$$e = e^2 \in RG \Rightarrow e \in R.$$

However this is false as seen by the following example.

Let $R = \mathbb{Z}_{(p)} \oplus \mathbb{Z}/p\mathbb{Z}$ where $\mathbb{Z}_{(p)}$ is the ring of those rational numbers whose denominators are powers of a given rational prime p and let $\langle h \rangle$ be a cyclic group of order p. Then

$$e = (1/p,0) \quad (1 + h + \cdots + h^{p-1})$$

is a nontrivial idempotent of $R\langle h \rangle$ and $p \notin \mathcal{U}R$, $e \notin R$. In view of this there exists the following conjecture, first made by Bovdi and Mihovski.

Conjecture 2.18. Let RG be the group ring of G over a commutative ring with identity R. Suppose that whenever G has an element of prime order p, $p \notin \mathcal{U}R$ and that R has no nontrivial idempotents. Then RG has no nontrivial idempotents.

We shall offer a proof of this conjecture in case G is finite or supersolvable and also if G is polycyclic-by-finite with R of characteristic 0.

Proposition 2.19. Suppose that $e = \Sigma e(g)g$ is an idempotent of RG where R is a ring with identity, without nontrivial idempotents, contained in a direct sum of fields,

$$R \subset F_1 \oplus \cdots \oplus F_k \oplus \cdots \oplus F_r$$

where F_i are of characteristic 0 for $1 \leq i \leq k \leq r$ and F_{k+j} for $j > 0$ are of nonzero characteristic. If G is polycyclic-by-finite then

$$e(1) = (\beta_1, \beta_2, \cdots, \beta_k, \cdots, \beta_r)$$

where $\beta_1 = \beta_2 = \cdots = \beta_k$.

Proof: We use induction on the Hirsch number $h(G)$. In view of (2.12) and (2.8) part (i) we have to prove the proposition only for finite groups. Suppose that $e(1) \neq 0$ and let

$$\rho: \left(\sum^{\oplus} F_i\right) G \to \left(\sum^{\oplus} F_i\right)_n, \quad n = |G|$$

be the regular representation of $(\Sigma^{\oplus} F_i)G$ with $\{g: g \in G\}$ as a basis of $(\Sigma^{\oplus} F_i)G$ over $\Sigma^{\oplus} F_i$, such that $\rho(e)$ is an idempotent matrix of degree n over R. Let us denote by $\chi(t)$ and $\chi_i(t)$ the characteristic polynomials of $\rho(e)$ and $\rho(\pi_i e) = \rho_i(e)$ respectively, where π_i is the natural projection $\Sigma_j^{\oplus} F_j \to F_i$. We have

$$\chi(t) = \det\left(tI - \rho(e)\right) = t^n - a_1 t^{n-1} + \cdots + (-1)^n a_n$$

with coefficients $a_i \in R$ such that for $1 \leq i \leq r$,

$$\chi_i(t) = \det\left(tI - \rho_i(e)\right) = t^n - \pi_i(a_1)t^{n-1} + \cdots + (-1)^n \pi_i(a_n).$$

On the other hand, we also know that

$$\chi_i(t) = t^{n-r_i}(t-1)^{r_i}$$

where r_i is the rank of the idempotent matrix $\rho_i(e)$ of degree n over F_i. Let m be the maximum of the ranks r_i. It follows that for i with $r_i = m$,

$$\chi_i(t) = t^{n-m}(t-1)^m = t^n - mt^{n-1} + \cdots + (-1)^m t^{n-m}.$$

Therefore, a_m is an idempotent with as many components equal to 1 as there are solutions to the equation $r_i = m$. Hence it follows that

$$a_m = (1,1,\cdots,1).$$

Thus the trace of $\rho(e) = ne(1) = (m,m,\cdots,m)$, proving the proposition.

<u>Theorem 2.20</u> (Cliff and Sehgal [1]). Let RG be the group ring of a polycyclic-by-finite group G over a commutative ring R with identity having characteristic 0. Suppose that $\mathcal{U}R \cap \{o(g): g \in G\} = \{1\}$ and that R has no nontrivial idempotents. Then RG has no nontrivial idempotents.

<u>Proof</u>: Let $e = \Sigma e(g)g \in RG$ be an idempotent. Replacing R by $\mathbb{Z}[e(g): g \in G]$ we may assume that R is Noetherian. Then replacing R by R/N where N is the ideal of nilpotent elements of R we may suppose that R has no nilpotent elements. This we may do since any idempotent of R/N can be lifted to that of R and since $\mathcal{U}R \cap \{o(g): g \in G\} = \mathcal{U}(R/N) \cap \{o(g): g \in G\}$. Since $\{0\}$ is the intersection of a finite number of prime ideals of R (or by using Goldie's Theorem on semi-prime rings), we have that

$$R \subseteq F_1 \oplus \cdots \oplus F_k \oplus \cdots \oplus F_r ,$$

a direct sum of fields where $0 \leq k \leq r$ and the F_i are of characteristic 0 for $1 \leq i \leq k$ whereas the F_{k+j} are of finite characteristic for $k > 0$. Since R has characteristic 0, certainly, $k \geq 1$. By considering $1 - e$ if necessary we may suppose that $e(1) \neq 0$. By the last proposition we have

$$e(1) = (m/n,m/n,\cdots,m/n,\alpha_1,\alpha_2,\cdots), \quad m/n \in Q, \quad \alpha_i \in F_{k+i} ,$$

where by the Theorem of Zalesskii (1.2), the α_i belong to finite fields. If $k = r$ we are finished by Kaplansky's Theorem (1.3), so we may suppose that $k < r$. Since $e(1) \neq 0$, it follows that $m \neq 0$ as otherwise a suitable power of $e(1)$ is a nontrivial idempotent in R. Let us write

$$n = ds, \quad m = dt, \quad (s,t) = 1,$$

and

$$e(1) = (t/s, t/s, \cdots, t/s, \alpha_1, \alpha_2, \cdots).$$

Since $(s,t) = 1$, there exist integers a and b such that $as + bt = 1$. Then

$$\beta = ae(1) + b1_R = (1/s, 1/s, \cdots, 1/s, a\alpha_1 + b, a\alpha_2 + b, \cdots) \in R.$$

We may suppose that $a\alpha_i + b \neq 0$ for any i, as otherwise a suitable power of $s\beta$ is a nontrivial idempotent in R. Now

$$s\beta - 1 = \left(0, 0, \cdots, 0, s(a\alpha_1 + b) - 1, \quad s(a\alpha_2 + b) - 1, \cdots\right) \in R.$$

Again, by the same argument,

$$s(a\alpha_i + b) - 1 = 0 \quad \text{for all } i,$$

and so, $a\alpha_i + b = 1/s$. We have

$$\beta = (1/s, 1/s, \cdots, 1/s) = \frac{1}{s} \cdot 1_R \in R.$$

Thus $s \in \mathcal{U}R$ and by Theorem 2.15, $s = 1$. Consequently by (1.3)

$$e(1) = (1, 1, \cdots, 1, \alpha_1, \alpha_2, \cdots).$$

Write $e' = 1 - e$. Then since $\pi_i(e')$, the projection of e' on F_iG, $1 \leq i \leq k$, has first coefficient 0, it follows that $\pi_i(e') = 0$. Then $e' \in SG$ where

$$S = F_{k+1} \oplus \cdots \oplus F_r.$$

Let I be the ideal of R generated by the coefficients of e'. Then $I = I^2 \subseteq S$. We claim:

(Krull) There exists an element $\gamma \in I$ such that $I(1-\gamma) = 0$.

Let I be generated as an R-module by $x_1, \cdots, x_n$. Then we can write

$$x_i = \sum_j y_{ij}x_j, \quad y_{ij} \in R, \quad 1 \le i \le n.$$

Therefore,

$$\sum_j (\delta_{ij}-y_{ij})x_j = 0$$

where δ_{ij} is the Kronecker delta. It follows that $dx_j = 0$ for all j if d is the determinant $|\delta_{ij}-y_{ij}|$. Since $(d-1) \in R$ the claim is proved.

Now $\gamma(\gamma-1) = 0$ implies that γ is an idempotent and thus $\gamma = 0$ or 1. Since $1 \notin S$, $\gamma \neq 1$ and it follows that $\gamma = 0$. Therefore $I = 0$. Consequently, $e' = 0$, $e = 1$. □

The next proposition was proved for integral domains R by Donald Coleman [3]. We need the well known lemma below.

Lemma 2.21. Let G be any group and K a commutative ring. Then the augmentation ideal, $\Delta(G)$ of KG, is nilpotent if and only if G is a finite p-group and K has characteristic p^ν for some prime p.

Proof: First suppose that $\Delta(G)$ is nilpotent. Suppose that G has two non-identity elements x and y of relatively prime orders m and n respectively. Then

$$1 = x^m = (x-1+1)^m = 1 + m(x-1) + \binom{m}{2}(x-1)^2 + \cdots + (x-1)^m$$

and

$$1 = y^n = (y-1+1)^n = 1 + n(y-1) + \binom{n}{2}(y-1)^2 + \cdots + (y-1)^n.$$

Choose integers u, v such that $mu + nv = -1$. Multiplying the above equations by $u(y-1)$ and $v(x-1)$ respectively and adding we obtain $(x-1)(y-1)$

$$= u\binom{m}{2}(x-1)^2(y-1)+\cdots+u(x-1)^m(y-1)+v\binom{n}{2}(x-1)(y-1)^2+\cdots+v(x-1)(y-1)^n$$

$$\in \Delta(G)(x-1)(y-1)+(x-1)(y-1)\Delta(G).$$

It follows that $(x-1)(y-1) \in \Delta^s(G)$ for all s. The element $(x-1)(y-1)$ is nonzero due to the choice of x and y. This is a contradiction. Thus every torsion element of G has p-power order.

Suppose that G is infinite. Then if $\Delta^n(G) = 0$, $\Delta^{n-1}(G) \neq 0$, pick $0 \neq \delta \in \Delta^{n-1}(G)$. We have $(g-1)\delta = 0 \Rightarrow g\delta = g$ for all $g \in G$. This is a contradiction since G is infinite. We have proved that G is a finite p-group.

If χ, the characteristic of R, is zero or has a prime divisor $q \neq p$ then $R \supseteq \mathbb{Z}/\chi\mathbb{Z}$. Also, there is a homomorphism $\mathbb{Z}/\chi\mathbb{Z} \to \mathbb{Z}/q\mathbb{Z}$ which induces a homomorphism $(\mathbb{Z}/\chi\mathbb{Z})G \to (\mathbb{Z}/q\mathbb{Z})G$. This implies that the augmentation ideal of $(\mathbb{Z}/q\mathbb{Z})G$ is nilpotent, in contradiction to Maschke's theorem. Thus $\chi = p^\nu$.

Conversely, suppose that G is a finite p-group and K has characteristic p^ν. We wish to prove that $\Delta(G)$ is nilpotent. We use induction on the order of G; the result being trivial if G is of order p as

$$\alpha \in \Delta(G) \Rightarrow \alpha^p = p\beta \quad \text{for some} \quad \beta \in KG \Rightarrow \alpha^{p\nu} = 0.$$

Let $N = \langle x \rangle$ be a subgroup of order p contained in the center of G. Then by induction we have $\Delta^k(G/N) = 0$ for some k. Thus

$$\Delta^k(G) \subseteq \Delta(G,N) = \Delta(N)KG = (1-x)KG$$

and hence

$$\Delta^{kp\nu}(G) = 0. \quad \square$$

<u>Proposition 2.22</u> (Sehgal-Zassenhaus). Let RG be the group ring of a finite group G over a commutative ring R. Suppose that R has no nontrivial idempotents and that every prime divisor of $|G|$ is a non-unit of R. Then RG has no nontrivial idempotents.

<u>Proof:</u> For characteristic 0 this proposition is a special case of the last one. We may therefore suppose that R has characteristic $m > 0$. Clearly m is a prime power p^ν as otherwise writing $m = m_1 m_2$, $(m_1, m_2) = 1$, R can be decomposed into a

direct sum of rings,

$$R = R_1 \oplus R_2, \quad \text{characteristic of } R_i = m_i.$$

This is not possible since R has no idempotents. So R has characteristic p^ν and G is a finite p-group because any natural number prime to p is invertible in R. Suppose that $e = \Sigma e(g)g = e^2 \in RG$. Then

$$\left(\sum e(g)\right)^2 = \left(\sum e(g)\right) \Rightarrow \sum e(g) = 0 \quad \text{or} \quad 1.$$

Therefore, e or $1 - e \in \Delta(G)$ which is nilpotent by (2.21). Hence $e = 0$ or 1. □

3. SUPERSOLVABLE GROUPS

We recall a few definitions and fix notation. A group G is said to be supersolvable if it has a normal series with cyclic factors:

$$G = G_1 \supseteq G_2 \supseteq \cdots \supseteq G_{n+1} = \{1\},$$

$$G_i \triangleleft G \quad \text{and} \quad G_i/G_{i+1} \quad \text{cyclic for} \quad 1 \le i \le n.$$

We define inductively subgroups $\zeta_i(G)$ and $\gamma_i(G)$ of a given group G. Let $\zeta_0(G) = \{1\}$ and let $\zeta_i(G)$ be given by

$$\zeta_i(G)/\zeta_{i-1}(G) = \zeta\left(G/\zeta_{i-1}(G)\right),$$

the center of $G/\zeta_{i-1}(G)$, for each $i \ge 1$. Let $\gamma_1(G) = G$ and let for each $i > 1$, $\gamma_i(G) = \left(\gamma_{i-1}(G),G\right)$, the group generated by all commutators $(x,y) = x^{-1}y^{-1}xy$, $x \in \gamma_{i-1}(G)$, $y \in G$.

The series

$$1 = \zeta_0(G) \subseteq \zeta_1(G) \subseteq \zeta_2(G) \subseteq \cdots \subseteq \zeta_n(G) \subseteq \cdots$$

and

$$G = \gamma_1(G) \supseteq \gamma_2(G) \supseteq \cdots \supseteq \gamma_n(G) \supseteq \cdots$$

are called the upper central series and lower central series respectively, of G. Recall that a central series in a group G is a normal series

$$G = G_1 \supseteq G_2 \supseteq G_3 \supseteq \cdots \supseteq G_{n-1} \supseteq G_n \supseteq \cdots$$

such that $G_i \lhd G$ and G_{i-1}/G_i is central in G/G_i for all i. A group G is said to be nilpotent if it has a finite central series and then both the lower and upper central series have the same length (see M. Hall [1]). For elements x, y, z, x_i of a group G, let us fix the following notations.

$$x^y = y^{-1}xy, \quad (x,y) = x^{-1}y^{-1}xy, \quad (x_1, x_2, \cdots, x_n) = \big((x_1, \cdots, x_{n-1}), x_n\big)$$

for all $n \geq 2$.

The following commutator identities are well known (see for example, M. Hall [1]).

Lemma 3.1. (i) $x^y = x(x,y)$;

(ii) $(xy,z) = (x,z)^y(y,z) = (x,z)(x,z,y)(y,z)$;

(iii) $(x,yz) = (x,z)(x,y)^z = (x,z)(x,y)(x,y,z)$;

(iv) $(x^{-1},y) = \big((x,y)^{-1}\big)^{x^{-1}}$;

(v) $(x,y^{-1}) = \big((x,y)^{-1}\big)^{y^{-1}}$.

Proposition 3.2. The derived group $G' = (G,G)$ of a supersolvable group G is nilpotent.

Proof: Suppose that $G = G_1 \supset G_2 \supset \cdots \supset G_{n+1} = \{1\}$ is a normal series for G with $G_i \lhd G$ and G_i/G_{i+1} cyclic for all $1 \leq i \leq n$. Write $H_i = G' \cap G_i$. Then since H_i/H_{i+1} is isomorphic to a subgroup of G_i/G_{i+1} we have after dropping repeated terms

$$G' = K_1 \supset K_2 \supset \cdots \supset K_s = \{1\}$$

where $K_i \lhd G$ and K_i/K_{i+1} cyclic for each i. We claim that this is a central series for G'. We have to show that K_i/K_{i+1} is central in G'/K_{i+1}. Every element of G induces by conjugation an automorphism of the cyclic group K_i/K_{i+1}. But since the automorphism group of a cyclic group is abelian, $(x,y) = x^{-1}y^{-1}xy$, $x,y \in G$, induces the identity automorphism on K_i/K_{i+1}. This says that K_i/K_{i+1} is central in G'/G_{i+1}. Hence G' is nilpotent.

Proposition 3.3. A supersolvable group G has a normal series

$$G \supset K = K_1 \supset K_2 \supset \cdots \supset K_s = M \supset \{1\}$$

with $K_i \lhd G$ where $K_s = M$ is a group of odd order, G/K is a 2-group and K_i/K_{i+1} are infinite cyclic.

Proof: We know that G has a normal series

(3.4) $$G = G_1 \supset G_2 \supset \cdots \supset G_{n+1} = \{1\},$$

with each $G_i \lhd G$ and G_i/G_{i+1} cyclic. Let $A \supset B \supset C$ be three successive terms in (3.4). It suffices to prove the following two statements.

(3.5) If A/B is odd cyclic and B/C is cyclic of infinite or 2-power order then we can move the odd factor to the right with the left factor infinite cyclic or a 2-group, i.e. we can find a normal subgroup D of G such that $A \supset D \supset C$ with A/D infinite cyclic or a 2-group and $|D/C|$ odd.

(3.6) If A/B is infinite cyclic and B/C is cyclic of 2-power order then we can move the infinite cyclic factor to the right with the left factor a 2-group.

By considering G/C we may assume that $C = \{1\}$, $B = \langle y \rangle$, $A = \langle x,y \rangle$. Since the automorphism group of $\langle y \rangle$ is a 2-group, x^{2^v} commutes with y for some $v \in \mathbb{Z}$.

Under the hypothesis of (3.5) x has odd order mod B so x commutes with y and $A = \langle x,y \rangle$ is Abelian. We have $A = E \times D$

where D is the subgroup of elements of odd order of A and as such is normal in G, and E is infinite cyclic or a 2-group and $A \supset D \supset \{1\}$ is the required series. Under the hypothesis of (3.6), $2^v \le |B|$ so $x^{|B|}$ commutes with y. Then $\langle x^{|B|}, y\rangle$ is an Abelian normal subgroup of G and

$$B = \langle x,y\rangle \supset \langle x^{|B|},y\rangle \supset \langle x^{|B|},y\rangle^{|B|} = \langle x^{|B|^2}\rangle \supset \{1\}$$

is the desired series of normal subgroups of G. This proves the proposition.

Proposition 3.7. Let G be a group with upper central series

$$\{1\} = \zeta_0(G) \subseteq \zeta_1(G) \subseteq \cdots \subseteq \zeta_n(G) \subseteq \cdots .$$

Suppose that $\zeta_1(G) = \zeta(G)$, the centre of G, is Π-free, for a set of rational primes Π i.e.

$$g \in \zeta(G), \quad g^{p_1 p_2 \cdots p_r} = 1, \quad p_i \in \Pi \Rightarrow g = 1.$$

Then each factor $\zeta_{i+1}(G)/\zeta_i(G)$ is Π-free.

Proof: We use induction on i. Suppose that $\zeta_1(G)/\zeta_0(G), \cdots, \zeta_i(G)/\zeta_{i-1}(G)$ are Π-free. Let $x \in \zeta_{i+1}(G)$ such that $x^m \in \zeta_i(G)$ where m is a product of primes in Π. Let $g \in G$ be any element. Then $(g,x) \in \zeta_i(G)$ and by the formula

$$(x,yz) = (x,z)(x,y)^z, \quad a^z = z^{-1}az,$$

we have that

$$(g,x)^m \equiv (g,x^m) \equiv 1 \bmod \zeta_{i-1}(G).$$

Therefore,

$$(g,x) \in \zeta_{i-1}(G) \Rightarrow x \in \zeta_i(G).$$

Consequently, $\zeta_{i+1}(G)/\zeta_i(G)$ is Π-free, completing the proof of the proposition.

We shall use the following special case.

Corollary 3.8. If G is a torsion free nilpotent group then its upper central factors $\zeta_{i+1}(G)/\zeta_i(G)$ are also torsion free.

Proposition 3.9. Let $G = \langle a_1, \cdots, a_r \rangle$ be a finitely generated group. Then for each $i \geq 1$, $\gamma_i(G)/\gamma_{i+1}(G)$ is generated by the finite set of elements $(b_1, \cdots, b_i)\gamma_{i+1}(G)$ where $b_j \in \{a_1, \cdots, a_r\}$. Here, for $i = 1$, $(b_1)\gamma_2(G)$ stands for $b_1\gamma_2(G)$.

Proof: We use induction on i and thus may assume that $i > 1$ and $\gamma_{i-1}(G)/\gamma_i(G)$ is generated by all elements $(b_1, \cdots, b_{i-1})\gamma_i(G)$, $b_j \in \{a_1, \cdots, a_r\}$. Now, $\gamma_i(G)/\gamma_{i+1}(G)$ is generated by

$$\{(y,x)\gamma_{i+1}(G) : y \in \gamma_{i-1}(G), \quad x \in G\}$$

$$= \Big\{(cz,x)\gamma_{i+1}(G) : c = c_1c_2\cdots c_t, \quad z \in \gamma_i(G), \quad x \in G, \quad \text{each}$$

$$c_j = (b_1, b_2, \cdots, b_{i-1})^{\varepsilon}, \quad \varepsilon = \pm 1\Big\}.$$

By Lemma 3.1, we have

$$(cz,x) = (c,x)(c,x,z)(z,x).$$

Because (c,x,z) and (z,x) lie in $\gamma_{i+1}(G)$, we have

$$(cz,x) \equiv (c,x) \bmod \gamma_{i+1}(G).$$

Thus the elements (c,x), c, x as above generate $\gamma_i(G)$ modulo $\gamma_{i+1}(G)$. Again by the same argument

$$(c,x) \equiv (c_1,x)(c_2,x) \cdots (c_t,x) \bmod \gamma_{i+1}(G).$$

Notice that, by Lemma 3.1,

$$(c_j,x) = \big((c_j^{-1},x)^{-1}\big)^{c_j} = (c_j^{-1},x)^{-1}\big((c_j^{-1},x)^{-1},c_j\big)$$

$$\equiv (c_j^{-1},x)^{-1} \bmod \gamma_{i+1}(G).$$

Thus we can conclude that the elements $(b_1,b_2,\cdots,b_{i-1},x)\gamma_{i+1}(G)$ generate $\gamma_i(G)/\gamma_{i+1}(G)$. Let $x = x_1x_2\cdots x_s$, each $x_j = a_\lambda$ or a_λ^{-1}. Write $b_0 = (b_1,\cdots,b_{i-1})$. Then

$$(b_0,x_1x_2\cdots x_s) \equiv (b_0,x_1)(b_0,x_2)\cdots(b_0,x_s) \bmod \gamma_{i+1}(G)$$

and

$$(b_0,a_\lambda^{-1}) \equiv (b_0,a_\lambda)^{-1} \bmod \gamma_{i+1}(G).$$

Hence, $\gamma_i(G)/\gamma_{i+1}(G)$ is generated by the elements

$$(b_1,b_2,\cdots,b_{i-1},a_\lambda)\gamma_{i+1}(G). \ \square$$

The next corollary may also be deduced from (3.9) and (2.2).

Corollary 3.10. Every subgroup of a finitely generated nilpotent group is finitely generated.

Proof: Let H be a subgroup of the finitely generated nilpotent group G, with lower central series

$$G = \gamma_1(G) \supset \gamma_2(G) \supset \cdots \supset \gamma_s(G) = \{1\}.$$

Setting $H_i = H \cap \gamma_i(G)$, we have

$$H = H_1 \supseteq H_2 \supseteq \cdots \supseteq H_s = \{1\},$$

where

$$\begin{aligned} H_i/H_{i+1} &= H \cap \gamma_i(G)/H \cap \gamma_{i+1}(G) \\ &\simeq \big(H \cap \gamma_i(G)\big)\gamma_{i+1}(G)/\gamma_{i+1}(G) \subseteq \gamma_i(G)/\gamma_{i+1}(G). \end{aligned}$$

Due to the last proposition, H_i/H_{i+1} is a subgroup of a finitely generated Abelian group. Hence H_i/H_{i+1} and consequently H is finitely generated. $\square$

Corollary 3.11. The upper central factors $\zeta_{i+1}(G)/\zeta_i(G)$ of a finitely generated torsion free nilpotent group G are finitely generated torsion free.

Proof: Corollaries 3.8 and 3.10.

Now we are able to prove Conjecture 2.18 for supersolvable groups.

Theorem 3.12 (Parmenter and Sehgal [1],[8]). Let RG be the group ring of a supersolvable group G over a commutative ring with identity R. Suppose that R has no nontrivial idempotents and that if G has an element of prime order p then $p \notin \mathcal{U}R$. Then RG has no nontrivial idempotents.

Lemma 3.13. Let N be a normal subgroup of G and R a commutative ring with identity. Then

$$\bigcap_n \Delta(G,N)^n = \left(\bigcap_n \Delta(N)^n\right)RG.$$

Proof: Let $\alpha \in \bigcap_n \Delta(G,N)^n$. Choosing a transversal T of N in G, we can express α, uniquely, as

$$\alpha = \sum \alpha_v t_v, \quad t_v \in T, \quad \alpha_v \in RN.$$

Since $\Delta(G,N)^n = \Delta(N)^n RG$, $\alpha_v \in \Delta(N)^n$ for each n, we have

$$\bigcap_n \Delta(G,N)^n \subseteq \left(\bigcap_n \Delta(N)^n\right)RG.$$

The opposite containment being trivial, the lemma is proved.

Lemma 3.14. Let R<x> be the group ring of an infinite cyclic group over a commutative ring with identity R. Then

$$\bigcap_n \Delta^n\langle x\rangle = 0.$$

Proof: Suppose $0 \neq \alpha \in \bigcap_n \Delta^n\langle x\rangle$. We may after multiplying α by a suitable power of x suppose that α is of the form $\Sigma_0^m a_i x^i$, $a_m \neq 0$, $m > 0$. Then $\alpha \notin \Delta^{m+1}\langle x\rangle$ because if it did we shall have an equation

$$\sum_0^m a_i x^i = (x-1)^{m+1}\beta, \quad \beta \in R\langle x\rangle,$$

which is seen to be impossible by comparing coefficients of the highest and the lowest degree terms. □

If we write Δ^ω for $\bigcap_n \Delta^n$, we may write $\Delta^{\omega^2} = (\Delta^\omega)^\omega = \bigcap_n (\Delta^\omega)^n$. Thus by induction $\Delta^{\omega^{t+1}} = (\Delta^{\omega^t})^\omega = \bigcap_n (\Delta^{\omega^t})^n$. We have the following useful result.

Lemma 3.15. Let G by a poly(infinite) cyclic group i.e.

$$G = G_1 \triangleright G_2 \triangleright G_3 \triangleright \cdots \triangleright G_t = 1$$

with each G_i/G_{i+1} infinite cyclic. Then for any commutative ring R, the augmentation ideal of RG satisfies

$$\Delta^{\omega^t}(G) = 0.$$

Proof: By the last lemma, $\Delta^\omega(G) \subseteq \Delta(G_2)RG$ and therefore

$$\Delta^{\omega^t}(G) \subseteq \Delta^{\omega^{t-1}}(G_2)RG$$

which is zero by induction on t. □

3.16. Proof of Theorem 3.12

In case R has characteristic 0, the result is included in Theorem 2.20. So we may assume that R has characteristic $m > 0$. Since R has no nontrivial idempotents, m is a prime power p^v. By Proposition 3.3, G has a normal series

$$G \supset M = M_1 \supset M_2 \supset \cdots \supset M_s = N \supset \{1\}$$

where each group is normal in G, N consists of all elements of odd order of G, G/M is a 2-group and M_i/M_{i+1} is infinite cyclic for each i. Let $e = \Sigma e(g)g = e^2 \in RG$. We may suppose that $e \in \Delta(G)$.

Case 1: p odd. In this case G has no 2-element and $H = G/N$ is torsion free supersolvable. By (3.2), H' is torsion free finitely generated nilpotent. So by (3.11) H' has a series

$$H' = H_1 \supset H_2 \supset \cdots \supset H_t = \{1\}$$

with H_i/H_{i+1} infinite cyclic. Let us write $\bar{e}$ for the image of e under the canonical map $RG \to R(G/N) = RH$. We know by Theorem 2.8 that $\tilde{\bar{e}}(h) = 0$ for all $h \in H$ and thus

$$\bar{e} \in \Delta(H,H') = \Delta(H')RH.$$

Moreover,

$$\bar{e} \in \Delta^{\omega^t}(H,H') = \Delta^{\omega^t}(H')RH$$

which is zero by (3.15). Hence $\bar{e} = 0$ i.e.

$$e \in \Delta(G,N) = \Delta(N)RG$$

which is nilpotent by (2.2) since N is a p-group. Hence $e = 0$.

Case 2: p = 2. In this case $N = 1$ as every odd integer is invertible in R. Since $\Delta(G/M)$ is nilpotent, $e \in \Delta(G,M) = \Delta(M)RG$. Again by (3.15), $e = 0$. □

4. CENTRAL IDEMPOTENTS

Now we shall prove a well known theorem of Schur (see P. Hall [1] or Robinson [1]). This theorem will be used in this section and also later on. We need the following

Proposition 4.1. Let H be a subgroup of index n in a group G generated by m elements. Then H is generated by at most $2nm$ elements.

Proof: Let us write $G = \langle x_1, \cdots, x_m \rangle$, $X = \{x_1, \cdots, x_m\}$, $X^{-1} = \{x_1^{-1}, \cdots, x_m^{-1}\}$. Pick a right transversal $T = \{1 = t_1, t_2, \cdots, t_n\}$

of H in G. So for $t_j \in T$, $g \in G$ we may express

$$t_j g = h(j,g)t_{(j)g}$$

for some $h(j,g) \in H$ and $t_{(j)g} \in T$. We notice that for $g,g' \in G$

$$t_j gg' = h(j,g)t_{(j)g}g' = h(j,g)h\big((j)g,g'\big)t_{((j)g)g'} \cdot$$

This implies that $t_{((j)g)g'} = t_{(j)gg'}$. Now every element $h \in H$ can be written as $h = y_1 y_2 \cdots y_r$, $y_i \in X \cup X^{-1}$. Thus

$$\begin{aligned} h = t_1 h &= h(1,y_1)t_{(1)y_1} \cdot y_2 \cdots y_r \\ &= h(1,y_1) \cdot h\big((1)y_1,y_2\big) \cdots h\big((1)y_1 y_2 \cdots y_{r-1},y_r\big)t_{(1)h} \\ &= h(1,y_1) \cdot h\big((1)y_1,y_2\big) \cdots h\big((1)y_1 y_2 \cdots y_{r-1},y_r\big). \end{aligned}$$

We have proved that H is generated by $\{h(j,y_i),\ y_i \in X \cup X^{-1},\ j = 1,\cdots,n\}$. □

Let us recall the definition of the transfer map. Let H be a subgroup of finite index n in G. Let T be a right transversal of H in G. For $g \in G$, $t_i \in T$ write

$$t_i g = h(i,g)t_{(i)g}, \quad h(i,g) \in H, \quad t_{(i)g} \in T.$$

The transfer of G into H is a mapping $\phi\colon G \to H/H'$ given by

$$\phi(g) = \prod_{i=1}^{n} t_i g\, t_{(i)g}^{-1} H' = \prod_{i=1}^{n} h(i,g)H'.$$

It is well known that ϕ is independent of the choice of T and is a homomorphism (see M. Hall [1]).

As a special case suppose that H is a subgroup of the center $\zeta(G)$ of G. Since for a fixed g, $t_i \to t_{(i)g}$ is a permutation of T we may renumber the t's in the following way.

$$t_1 g = h(1,g)t_2,\ t_2 g = h(2,g)t_3, \cdots,$$
$$t_{n_1} g = h(n_1,g)t_1; \cdots ; \cdots t_{n_{s+1}} g = h(n_{s+1},g)t_{n_s+1} \cdot$$

Then the transfer map is easy to compute.

$$\begin{aligned}\phi(g) &= \left(t_1 g t_2^{-1} \cdot t_2 g t_3^{-1} \cdots t_{n_1} g t_1^{-1}\right) \cdots \\ &= t_1 g^{n_1} t_1^{-1} \cdot t_{n_1+1} g^{n_2} t_{n_1+1}^{-1} \cdots t_{n_s} g^{n_s} t_{n_s}^{-1} \\ &= g^{n_1+n_2+\cdots+n_s} = g^n .\end{aligned}$$

Thus if H is central, the transfer map is given by $g \to g^n$.

Theorem 4.2 (Schur). If $(G: \zeta(G)) = n$ then G' is finite and $(G')^n = 1$.

Proof: As we have seen $\phi: G \to \zeta(G)$ is the map $g \to g^n$. Since G' is in the kernel of ϕ we have $(G')^n = 1$. It remains to prove that G' is finite. Since for $g_1, g_2 \in G$, $z_1, z_2 \in \zeta(G)$, $(g_1 z_1, g_2 z_2) = (g_1, g_2)$ it follows that G' is finitely generated. Now,

$$G'/G' \cap \zeta(G) \simeq G'\zeta(G)/\zeta(G)$$

is finite and therefore by (4.1), $G' \cap \zeta(G)$ is finitely generated. Thus $G' \cap \zeta(G)$ is finite. Hence G' is finite. □

Corollary 4.3. Let G be a solvable group satisfying $G^{p^m} \subseteq \zeta(G)$ for a fixed prime power p^m. Then $(G')^{p^M} = 1$ for a fixed M.

Proof: We use induction on the solvability length of G. Thus $(G'')^{p^\ell} = 1$ for a fixed ℓ. We may factor by G'' and assume that G is metabelian. Since G' is Abelian it is enough to prove that for all $x, y \in G$, $(xyx^{-1}y^{-1})^{p^M} = 1$ for a fixed M. Let $H = \langle x,y \rangle$. Then $|H/H'\zeta(H)| \leq p^{2m}$. Therefore it follows by (4.1) that $H'\zeta(H)$ is generated by at most $4 \cdot p^{2m}$ elements. Thus

$$|H'\zeta(H)/\zeta(H)| \leq p^{m \cdot 4p^{2m}}, \quad |H/\zeta(H)| \leq p^M, \quad M = 2m + 4mp^{2m}.$$

Hence by the last theorem $(H')^{p^M} = 1$. Since M is independent of H, the corollary follows. □

We shall need the following result of Burns [1] and Passman [3].

Theorem 4.4. Let $e = \Sigma e(g)g$ be a central idempotent of a group ring RG, of G over an arbitrary ring R with identity. Then the support group of e,

$$\langle \text{supp}(e) \rangle = \langle g : e(g) \neq 0 \rangle$$

is a finite normal subgroup of G.

Lemma 4.5 (Bass). Any idempotent of RG, the group ring of an Abelian group G, with torsion subgroup T, over a commutative ring R with identity belongs to RT.

Proof: We may suppose that G is finitely generated and

$$G = T \times \langle x_1 \rangle \times \cdots \times \langle x_k \rangle, \quad o(x_i) = \infty.$$

By induction on k it suffices to prove that for any Abelian group A,

$$e = e^2 \in R(A \times \langle x_1 \rangle) \Rightarrow e \in RA.$$

Now since $R(A \times \langle x_1 \rangle) = (RA)\langle x_1 \rangle$, we may write

$$e = \sum e(g)g, \quad e(g) \in RA, \quad g \in \langle x_1 \rangle.$$

Then $e_0 = \Sigma e(g) = e_0^2$ and $(e - e_0 e)$ and $(e_0 - e_0 e)$ are idempotents of $(RA)\langle x_1 \rangle$ lying in $\Delta_{RA}\langle x_1 \rangle$. Since by Lemma 3.14,

$$\bigcap_n \Delta_{RA}\langle x_1 \rangle^n = 0,$$

we conclude that $e - e_0 e = 0 = e_0 - e_0 e$ and consequently $e = e_0 \in RA$.

4.6. Proof of Theorem 4.4 (Burns)

Since e is central in RG, each e(g) lies in the center of R and we may assume that R is commutative. We may also suppose that G is a finitely generated FC-group i.e. each element of G has a finite number of conjugates; say, $G = \langle g_1, \cdots, g_r \rangle$. Then since $(G: C_G(g_i))$ is finite for each i, G is a finite extension of its center $\zeta(G)$. As a result, by (4.2) the derived group G' is finite. Thus the torsion elements of G form a normal subgroup G^+ of G with G/G^+ torsion free Abelian. Let us write the finitely generated Abelian group $\zeta(G)$ as

$$\zeta(G) = T \times F$$

where T is finite and F is torsion free. Then

$$\sigma: G \to G/F \times G/G^+$$

given by

$$g^\sigma = (gF, gG^+) \text{ is an imbedding, where,}$$

$$G/F \times G/G^+ = G^\sigma \cdot (\{1\} \times G/G^+).$$

Now, $e^\sigma = \Sigma e(g) g^\sigma$ is a central idempotent of RG^σ; actually, e^σ is central in $R(G/F \times G/G^+)$ as G/G^+ is central in $G/F \times G/G^+$. Writing

$$G/F = B, \quad G/G^+ = A,$$

we have e^σ a central idempotent of $R(B \times A) = (RB)(A)$. Thus $e^\sigma \in SA$ where S is a commutative subring of RB. Since A is torsion free Abelian, it follows by Lemma 4.5 that $e^\sigma \in S \subseteq RB$. But B is finite as $(G: \zeta(G))$ and $(\zeta(G): F) = |T|$ are both finite, hence the support group of e^σ and therefore e is finite. Normality of the support group is obvious from the centrality of e. □

<u>Definition</u>. A group G is said to be residually finite if, for every $g \in G$ there exists a normal subgroup N_g of G such that $g \notin N_g$ and G/N_g is finite.

<u>Lemma 4.7</u>. A finitely generated FC-group G is residually finite.

<u>Proof</u>: Let $G = \langle g_1, \cdots, g_n \rangle$. Then $(G: C(g_i)) < \infty$ for all i and consequently $(G: \zeta) = (G: \bigcap_i C(g_i)) < \infty$ where ζ is the center of G. It follows by (4.1) that ζ is finitely generated. Therefore there exists a natural number t such that $A = \zeta^t$ is torsion free finitely generated Abelian. We have

$$(G: A) < \infty, \quad A = \langle x_1 \rangle \times \cdots \times \langle x_\ell \rangle, \quad o(x_i) = \infty.$$

If $g \notin A$ then set $N_g = A$. If $g \in A$ then there exists a natural number m such that $g \notin A^m$ and we take $N_g = A^m$. □

CHAPTER II

UNITS IN GROUP RINGS I

1. UNITS OF FINITE ORDER

Let us recall from (I.1.2) and (I.1.3) that if e is a nontrivial idempotent of $\mathbb{C}G$, then $e(1) \in Q$ and $0 < e(1) < 1$.

<u>Theorem 1.1</u>. Let $\alpha = \Sigma\alpha(g)g$ be a unit of finite order, $\alpha^n = 1$, with $\alpha(1) \neq 0$ in the group ring RG where R is a subring of the complex number field $\mathbb{C}$. Then

(i) $\alpha = \sum_1^t \xi_i e_i$, $\xi_i \in \mathbb{C}$, $\xi_i^n = 1$, $\{e_i\}$ a set of nonzero commuting orthogonal idempotents of $\mathbb{C}G$;

(ii) Writing the nonzero rational numbers $e_i(1)$ as r_i/s_i, $r_i, s_i \in \mathbb{Z}$, $(r_i, s_i) = 1$, $r_i > 0$; if no prime divisor of any s_i is a unit of R then we have $\alpha = \alpha(1)$.

<u>Proof</u>: The commutative finite dimensional algebra $\mathbb{C}[\alpha]$ can be expressed as a direct sum of isomorphic copies of $\mathbb{C}$,

$$\mathbb{C}[\alpha] = \mathbb{C}e_1 \oplus \cdots \oplus \mathbb{C}e_t$$

where $1 = e_1 + \cdots + e_t$, $e_i e_j = \delta_{ij} e_j$ and δ_{ij} is the Kronecker delta. This can be seen by observing that $\mathbb{C}[\alpha]$ is isomorphic to $\mathbb{C}[x]/\langle f(x)\rangle$, the factor ring of the polynomial ring $\mathbb{C}[x]$ by a polynomial $f(x)$ which being a divisor of $x^n - 1$ has all distinct roots. We have then,

$$\alpha = \xi_1 e_1 + \cdots + \xi_t e_t, \quad \xi_i^n = 1$$

$$\alpha(1) = \xi_1 e_1(1) + \cdots + \xi_t e_t(1).$$

Therefore, for any natural number r,

$$\alpha^r = \xi_1^r e_1 + \cdots + \xi_t^r e_t \in RG.$$

Since any automorphism of the n-th cyclotomic field $K = Q(\xi)$, $\xi^n = 1$ is induced by $\xi \to \xi^r$ for suitable r, we have that any conjugate of $\alpha(1)$ being of the form

$$\xi_1^r e_1(1) + \cdots + \xi_t^r e_t(1)$$

equals $\alpha^r(1)$ and belongs to R. Thus the Norm from K to Q of $\alpha(1)$,

$$N_{K/Q}\alpha(1) = N\alpha(1) \in R.$$

We have,

$$\alpha(1) = \sum e_i(1)\xi_i = \sum r_i/s_i \xi_i, \quad r_i > 0, \quad \sum r_i/s_i = 1 \qquad (*)$$

$$s\alpha(1) = \sum r_i' \xi_i \quad s = \Pi s_i, \quad \sum r_i' = s. \qquad (**)$$

Taking Norm from K to Q of both sides we obtain,

$$0 \neq s^\ell N\alpha(1) = \prod_i \sum r_i' \xi_i ,$$

a sum of s^ℓ roots of unity, where $\ell = (K : Q)$. Since $s^\ell N\alpha(1)$ is a rational number as well as an algebraic integer, $s^\ell N\alpha(1) \in \mathbb{Z}$. But the primes involved in the factorization of s^ℓ are primes dividing s_i which are not invertible in R. Hence $N\alpha(1) \in \mathbb{Z}$. By taking the absolute value of both sides of (*) we see that $|\alpha(1)| \leq 1$ and that the same is true for every conjugate of $\alpha(1)$. Since $N\alpha(1)$ is a nonzero rational integer it follows that $|\alpha(1)| = 1$ and therefore all the ξ_i's in (*) are equal. Thus

$\alpha(1) = \xi_1$ and $\alpha = \Sigma\xi_i e_i = \xi_1$. □

Corollary 1.2 (Bass [2]). Let R be an integral domain of characteristic 0 such that no rational prime is a unit of R. Then

$$\alpha \in RG, \quad \alpha^n = 1, \quad \alpha(1) \neq 0 \Rightarrow \alpha = \alpha(1).$$

Corollary 1.3 (Passman). $\alpha \in \mathbb{Z}G$, $\alpha^n = 1$, $\alpha(1) \neq 0 \Rightarrow \alpha = \alpha(1)$.

This corollary was proved for finite groups by Berman [3] and Cohn and Livingstone [2].

Corollary 1.4. Let R be an integral domain of characteristic 0 and G a polycyclic-by-finite group such that $\mathcal{U}R \cap \{o(g): g \in G\} = \{1\}$. Then

$$\alpha \in RG, \quad \alpha^n = 1, \quad \alpha(1) \neq 0 \Rightarrow \alpha = \alpha(1).$$

Proof: (I.2.15).

In view of this we make the following conjecture which, of course, is a consequence of the Conjecture (I.2.11).

Conjecture 1.5. Let R be an integral domain of characteristic 0 satisfying for a group G, $\mathcal{U}R \cap \{o(g): g \in G\} = \{1\}$. Then

$$\alpha \in RG, \quad \alpha^n = 1, \quad \alpha(1) \neq 0 \Rightarrow \alpha = \alpha(1).$$

For $\alpha = \Sigma\alpha(g)g \in \mathbb{C}G$, define $\alpha^* = \Sigma\overline{a(g)}g^{-1}$, where $^{-}$ denotes the complex conjugate. Then it is easy to see that for $a \in \mathbb{C}$ and $\alpha, \beta \in \mathbb{C}G$ we have

(1) $(\alpha^*)^* = \alpha$
(2) $(\alpha+\beta)^* = \alpha^* + \beta^*$
(3) $(a\alpha)^* = \bar{a}\alpha^*$, and
(4) $(\alpha\beta)^* = \beta^*\alpha^*$.

Corollary 1.6. Let α be a unit of finite order in $\mathbb{Z}G$ such that $\alpha\alpha^*$ is also of finite order. Then $\alpha = \pm g$, $g \in G$.

Proof: Since $\alpha\alpha^* = \left(\Sigma\alpha(g)^2\right)\cdot 1 + \Sigma_{g\neq 1} a_g g$, $a_g \in \mathbb{Z}$, is a unit of finite order with $\Sigma\alpha(g)^2 \neq 0$, it follows by Corollary (1.3) that $\alpha\alpha^* = \Sigma_g \alpha(g)^2$, a natural number. Thus $\Sigma_g \alpha(g)^2 = 1$ so that $\alpha = \pm g$ for some $g \in G$. □

Corollary 1.7. Any central unit of finite order in $\mathbb{Z}G$ is of the form $\pm t$ where t is a torsion element of the center of G.

Corollary 1.8. Let G be an Abelian group. Then the torsion subgroup $T\mathcal{U}(\mathbb{Z}G)$ of the unit group, $\mathcal{U}(\mathbb{Z}G)$ is given by

$$T\mathcal{U}(\mathbb{Z}G) = \pm TG.$$

This last corollary is a phenomenon which rarely happens in non-commutative groups. In the next section we investigate it further. We shall call a unit of RG trivial if it is of the form ug, $u \in \mathcal{U}R$, $g \in G$.

2. $\mathbb{Z}G$ WITH TRIVIAL UNITS OF FINITE ORDER

Proposition 2.1 (G. Higman). Let $G^* = G \times \langle x\rangle$, $x^2 = 1$. Then

$$\mathcal{U}(\mathbb{Z}G) = \pm G \Rightarrow \mathcal{U}(\mathbb{Z}G^*) = \pm G^*.$$

Proof: Let $\alpha,\beta,\gamma,\delta \in \mathbb{Z}G$ be such that $(\alpha+\beta x)(\gamma+\delta x) = 1$. It suffices to prove that $\alpha = 0$ or $\beta = 0$. By comparing coefficients we have

$$\alpha\gamma + \beta\delta = 1, \quad \alpha\delta + \beta\gamma = 0.$$

Therefore,

$$(\alpha+\beta)(\gamma+\delta) = 1, \quad (\alpha-\beta)(\gamma-\delta) = 1.$$

By assumption on $\mathcal{U}(\mathbb{Z}G)$

$$\alpha + \beta = \pm g_1, \quad \alpha - \beta = \pm g_2, \quad g_1, g_2 \in G.$$

Thus $2\alpha = \pm g_1 \pm g_2$. The only way for $\pm g_1 \pm g_2$ to have even coefficients is for g_1 to be equal to g_2. It follows that $\alpha = 0$ or $\beta = 0$. In any case, the unit $\alpha + \beta x$ is trivial. □

Proposition 2.2 (G. Higman). Let G be the Quaternion group of order 8. Then $\mathcal{U}(\mathbb{Z}G) = \pm G$.

Proof: The group G is given by $\langle x,y: x^4 = y^4 = 1,\ x^2 = y^2,\ yx = x^{-1}y\rangle$. Writing $xy = z$, let $\alpha = a_0 + a_1x + a_2y + a_3z + (b_0+b_1x+b_2y+b_3z)x^2$ be a unit of $\mathbb{Z}G$. Consider the homomorphism of $\mathbb{Z}G$ to $\mathbb{Z}[1,i,j,k]$, the ring of integral quaternions, induced by $x \to i$, $y \to j$, $xy \to k$. The image of α is $(a_0-b_0) + (a_1-b_1)i + (a_2-b_2)j + (a_3-b_3)k$. Since the only units of $\mathbb{Z}[1,i,j,k]$ are ± 1, $\pm i$, $\pm j$ and $\pm k$ we have

$$(2.3)\qquad \begin{aligned} &a_r - b_r = \pm 1 \quad \text{for some} \quad r = 0,1,2 \text{ or } 3 \quad \text{and} \\ &a_s - b_s = 0 \quad \text{for} \quad s \neq r. \end{aligned}$$

Again, let $H = \langle x'\rangle \times \langle y'\rangle$, $x'^2 = 1 = y'^2$. Consider the homomorphism $\mathbb{Z}G \to \mathbb{Z}H$ induced by $x \to x'$, $y \to y'$. The image of α is $(a_0+b_0) + (a_1+b_1)x' + (a_2+b_2)y' + (a_3+b_3)x'y'$. Since the units of $\mathbb{Z}H$ are trivial by Proposition 2.1 we have

$$(2.4)\qquad \begin{aligned} &a_\ell + b_\ell = \pm 1 \quad \text{for some} \quad \ell = 0,1,2 \text{ or } 3, \\ &a_m + b_m = 0 \quad \text{for} \quad m \neq \ell. \end{aligned}$$

Since a_i, b_i are integers it follows from (2.3) and (2.4) that $\ell = r$. Hence either

$$a_r = \pm 1, \quad b_r = 0, \quad a_s = b_s = 0 \quad \text{for} \quad s \neq r$$

or

$$a_r = 0, \quad b_r = \pm 1, \quad a_s = b_s = 0 \quad \text{for} \quad s \neq r.$$

In either case α is a trivial unit. □

Corollary 2.5. Let $G = A \times B$, where A is an elementary Abelian 2-group and B is the quaternion group of order 8. Then all units in $\mathbb{Z}G$ are trivial.

We shall need the next result in the following chapters also.

Proposition 2.6. Let G be a finite Abelian group and F a field whose characteristic does not divide $(G: 1)$. Then

$$FG \simeq \sum^{\oplus} a_d\, F(\xi_d), \tag{2.7}$$

where ξ_d is a primitive d-th root of unity and $a_d\big(F(\xi_d): F\big) = n_d$, the number of elements of order d in G. Here, $a_d F(\xi_d)$ denotes a sum of a_d copies of $F(\xi_d)$.

Proof: Let ψ be an irreducible representation of FG. Then by Schur's Lemma, $\psi(G)$ is a finite commutative group contained in a division ring and therefore is cyclic. Suppose $\psi(G) = \langle\psi(g)\rangle$ with some $g \in G$. Then the irreducible module associated with ψ is $F \cdot \psi(G) = F\langle\psi(g)\rangle \simeq F(\xi_d)$ if $\psi(G)$ is cyclic of order d. So we may write

$$FG = \sum^{\oplus} a_d\, F(\xi_d) \tag{2.8}$$

where $F(\xi_d)$ corresponds to a representation ψ with $\psi(G)$ cyclic of order d. We have to prove that $a_d\big(F(\xi_d): F\big) = n_d$.

Let $K = F(\xi_e)$, where e is the exponent of G. Then every irreducible representation ψ of G over K is linear and $G/\mathrm{Ker}\,\psi \simeq \psi(G)$ is cyclic say of order d. And every cyclic factor group of G of order d gives such an irreducible representation of KG. So the number of irreducible representations ψ of KG with $\psi(G)$ cyclic of order d is

$= \phi(d) \times$ (the number of cyclic factor groups of G of order d)
$= \phi(d) \times$ (the number of cyclic subgroups of G of order d)
$=$ the number of elements of order d in G
$= n_d$.

Hence,

$$KG = \overset{\oplus}{\sum} n_d\, K_d$$

where $K_d = K(\xi_d) = K$ corresponds to an irreducible representation ψ of G with $\psi(G)$ cyclic of order d.

Let ψ be an irreducible representation of FG corresponding to $F(\xi_d)$. Then there exists a $g \in G$ with $\psi(g)$ a matrix of multiplicative order d. Also, there exists an invertible matrix X with entries in K such that

$$X A X^{-1} = \begin{bmatrix} \xi_d & & \\ & \ddots & \\ & & \ddots \end{bmatrix} .$$

Applying any automorphism $\sigma \in \mathcal{G}$, the Galois group of $F(\xi_d)$ over F, we get that A is similar to the diagonal matrix whose diagonal terms are ξ_d^{σ}, $\sigma \in \mathcal{G}$. Thus ψ splits into $\left(F(\xi_d): F\right) = (\mathcal{G}: 1)$ irreducible representations over K each corresponding to K_d. Hence $n_d = a_d \left(F(\xi_d): F\right)$. □

Let G be a finite Abelian group. Then by the last proposition we may write

$$QG = \overset{\oplus}{\sum} Q(\xi_d),$$

a direct sum of cyclotomic fields $Q(\xi_d)$ where ξ_d is a primitive d-th root of unity. Clearly,

$$\mathbb{Z}G \subseteq \overset{\oplus}{\sum} \mathbb{Z}[\xi_d] = R.$$

We claim

(2.9) $(\mathcal{U}R: \mathcal{U}\mathbb{Z}G) < \infty.$

Proof: The additive groups of R and $\mathbb{Z}G$ are both free Abelian of rank $|G|$. Therefore, the index of additive groups,

$$(R: \mathbb{Z}G) = \ell < \infty.$$

Consequently, $\ell R \subset \mathbb{Z}G$. Let $x, y \in \mathcal{U}R$ and $x + \ell R = y + \ell R$, then $x^{-1}y - 1 \in \ell R \subseteq \mathbb{Z}G$ and thus $x^{-1}y \in \mathbb{Z}G$. Similarly $y^{-1}x \in \mathbb{Z}G$ and therefore $x^{-1}y \in \mathcal{U}\mathbb{Z}G$. It follows that

$$(\mathcal{U}R : \mathcal{U}\mathbb{Z}G) \leq (R : \ell R) < \infty. \quad \square$$

The following lemma is a well known consequence of Dirichlet's Unit Theorem (see Hasse [1], page 536).

Lemma 2.10. Let $F = Q(\xi)$, $\xi^n = 1$, $n > 2$ be the n-th cyclotomic field. Denote by F_0 the fixed field of the automorphism $\xi \to \xi^{-1}$. For any subfield K of F let O_K be its ring of algebraic integers. Then $\mathcal{U}O_K$ has the same rank as $\mathcal{U}O_F = \mathcal{U}\mathbb{Z}[\xi]$ if and only if $K = F$ or F_0.

Lemma 2.11. Let $\eta = \Sigma\eta(g)g \in RG$ be nilpotent, $\eta^n = 0$ where R is an integral domain of characteristic 0. Then $\eta(1) = 0$.

Proof: We may assume that R is finitely generated. Suppose that $\eta(1) \neq 0$. By (I.1.9), there exists a rational prime p such that the support of η has no element of p-power order $\neq 1$ and a homomorphism

$$\phi_p : R \to F_p, \quad \phi_p(\eta(1)) \neq 0,$$

where F_p is a field of characteristic p. Writing $\alpha = \Sigma\phi_p(\eta(g))g$, we have, for large enough m,

$$0 = \alpha^{p^m} = \sum \alpha(g)^{p^m} g^{p^m} + \beta$$

for some $\beta \in [F_pG, F_pG]$, by (I.1.4). It follows by (I.1.5) that

$$\alpha(1)^{p^m} = 0, \quad \alpha(1) = 0, \quad \phi_p(\eta(1)) = 0$$

which is a contradiction, proving the lemma. $\square$

Corollary 2.12. RG has no nilpotent ideals if R is an integral domain of characteristic 0.

Corollary 2.13. RG has no nilpotent elements in its centre if R is an integral domain of characteristic 0.

We wish to characterize all finite group G such that all units of finite order in $\mathbb{Z}G$ are trivial. The next proposition gives useful implications.

Proposition 2.14. Let G be any group. Then we have the implications

(a) $\Rightarrow$ (b) $\Rightarrow$ (c) if G is finite and (c) $\Rightarrow$ (d) $\Rightarrow$ (e)

for the following statements:

(a) The torsion units of $\mathbb{Z}G$ are trivial;

(b) $G \lhd \mathcal{U}(\mathbb{Z}G)$ i.e. G is a normal subgroup of $\mathcal{U}\mathbb{Z}G$;

(c) $(\mathcal{U}\mathbb{Z}G)^n \subseteq \zeta(\mathcal{U}\mathbb{Z}G)$ for some n i.e. $\mathcal{U} = \mathcal{U}\mathbb{Z}G$ is of bounded exponent over its center $\zeta\mathcal{U}$;

(d) $\mathbb{Z}G$ has no (nonzero) nilpotent elements;

(e) Every finite subgroup of G is normal.

Proof: (i) (a) $\Rightarrow$ (b) is clear as $\pm G \lhd \mathcal{U}\mathbb{Z}G$ and ugu^{-1} has augmentation one for every $u \in \mathcal{U}\mathbb{Z}G$, $g \in G$.

(ii) To prove (b) $\Rightarrow$ (c), let $u \in \mathcal{U}\mathbb{Z}G$. Since $G \lhd \mathcal{U}\mathbb{Z}G$, u gives an automorphism of G by conjugation. Let n be the order of the automorphism group of G. Then

$$u^n g u^{-n} = g \quad \text{for all} \quad g \in G.$$

Consequently, $u^n \in \zeta(\mathcal{U}\mathbb{Z}G)$, proving (c).

(iii) (c) $\Rightarrow$ (d): Suppose that we have $\mu \in \mathbb{Z}G$ with $\mu^2 = 0$. Then $(1+\mu)(1-\mu) = 1$, $(1+\mu) \in \mathcal{U}\mathbb{Z}G$ and consequently $(1+\mu)^n = 1 + n\mu \in \zeta(\mathcal{U}\mathbb{Z}G)$. It follows that $n\mu$, and thus μ, is a central element of $\mathbb{Z}G$. Hence $\mu = 0$ by (2.13).

(iv) (d) $\Rightarrow$ (e): Let $t \in G$, $t^m = 1$. Then for any $g \in G$,

$$\mu = (1-t)g(1+t+\cdots+t^{m-1})$$

satisfies $\mu^2 = 0$. Thus $\mu = 0$ and we have

$$g(1+t+\cdots+t^{m-1}) = tg(1+t+\cdots+t^{m-1}).$$

This implies a relation of the type $g = tgt^i$, proving the normality of $\langle t\rangle$ and thus (e). $\square$

The next result will also be used in Chapter VI to characterize groups G with $\mathcal{U}\mathbb{Z}G$ an FC-group. We call a group Hamiltonian if it is non-Abelian and all its subgroups are normal. It is well known (see M. Hall [1]) that a Hamiltonian group can be written as $A \times E \times K_8$ where K_8 is the quaternion group of order 8, A is Abelian with every element of odd order, and E is an elementary Abelian 2-group.

Theorem 2.15 (Sehgal-Zassenhaus [1]). Let G be a group satisfying $(\mathcal{U}\mathbb{Z}G)^n \subseteq \zeta(\mathcal{U}\mathbb{Z}G)$ for some n. Then the torsion elements TG of G form an Abelian or a Hamiltonian 2-group. Moreover, for any Abelian subgroup T_1 of T and $x \in G$ we have

(2.16) either x centralizes T_1 or $x^{-1}tx = t^{-1}$ for all $t \in T_1$.

Proof: We have a group G with $(\mathcal{U}\mathbb{Z}G)^n \subseteq \zeta(\mathcal{U}\mathbb{Z}G)$. So by the last proposition, every finite subgroup of G is normal and consequently $TG = T$ is an Abelian or a Hamiltonian group. We claim

(2.17) $x \in G,\ \ t \in T,\ \ x^{-1}tx \neq t \Rightarrow xtx^{-1} = t^{-1}.$

We know by (2.6) that

$$Q\langle t\rangle = \sum_{d|m}^{\oplus} F_d, \quad F_d = Q(\xi_d)$$

where ξ_d is a primitive d-th root of unity, m is the order of t. Also it is easy to see that

$$t = \sum \xi_d, \quad x^{-1}tx = t^i = \sum \xi_d^i, \quad (i,m) = 1.$$

Because $\xi_d \to \xi_d^i$ is an automorphism of $Q(\xi_d)$,

$$\sum_j a_j \xi_d^j = 0, \quad a_j \in Q \Rightarrow \sum a_j \xi_d^{ij} = 0$$

and thus $x^{-1}Q(\xi_d)x = Q(\xi_d)$, $x^{-1}\mathbb{Z}[\xi_d]x = \mathbb{Z}[\xi_d]$. Considering the group $\langle x,t \rangle$ and applying (2.9) and the hypothesis we conclude that $U_1 = (U\mathbb{Z}[\xi_m])^s$ commutes element wise with x for a suitable integer $s \geq 1$. Thus $Q(U_1) \subseteq \overline{F}_m$, the fixed field of F_m under the automorphism $f \to x^{-1}fx$, $f \in F_m$. But U_1 has the same rank as $U\mathbb{Z}[\xi_m]$ and thus by (2.10)

$$\overline{F}_m = F_m \quad \text{or} \quad (F_m : \overline{F}_m) = 2, \quad x^{-1}\xi_m x = \xi_m^{-1} .$$

In any case, $x^{-1}\xi_m x = \xi_m$ or ξ_m^{-1}. But $t = \Sigma_{d|m}\xi_d$, $x^{-1}tx = t^i$ and therefore $x^{-1}tx = t$ or t^{-1}, proving (2.17).

Let $t_1, t_2 \in T_1$, $x^{-1}t_1x = t_1$, $x^{-1}t_2x = t_2^{-1}$. Then $x^{-1}t_1t_2x = t_1t_2^{-1}$. It follows by (2.17) that either $t_1 = t_1^{-1}$ or t_2^{-1} proving (2.16). It remains to prove that if T is Hamiltonian then T has no nontrivial element of odd order. Suppose $T = A \times E \times K_8$ where A is Abelian with every element of odd order, E is an elementary Abelian 2-group and $K_8 = \langle i,j\colon i^2 = j^2 = t,\ t^2 = 1,\ ji = ijt \rangle$ is the quaternion group of order 8. Then for $a \in A$,

$$j^{-1}aj = a, \quad j^{-1}ij = i^{-1}.$$

Applying (2.16) with $x = j$ we have $a = a^{-1}$ and thus $a = 1$. We have proved that $T = E \times K_8$, a Hamiltonian 2-group, completing the proof of the theorem. □

We now have the following theorem, a part of which goes back to Berman [1]. Denote by TG the set of torsion elements of G.

Theorem 2.18. Let G be a finite group. Then

$T\mathcal{U}(\mathbb{Z}G) = \pm G \iff G \lhd \mathcal{U}\mathbb{Z}G \iff (\mathcal{U}\mathbb{Z}G)^n \subseteq \zeta(\mathcal{U}\mathbb{Z}G)$ for some $n \iff G$ is Abelian or a Hamiltonian 2-group.

Proof: We have from (2.14) that

$$T\mathcal{U}\mathbb{Z}G = \pm G \Rightarrow G \lhd \mathcal{U}\mathbb{Z}G \Rightarrow (\mathcal{U}\mathbb{Z}G)^n \subseteq \zeta(\mathcal{U}\mathbb{Z}G) \text{ for some } n,$$

which implies by (2.15) that G is Abelian or a Hamiltonian 2-group. If G is Abelian then, of course, by (1.8), $T\mathcal{U}\mathbb{Z}G = \pm G$ and if G is a Hamiltonian 2-group, then by (2.5) $\mathcal{U}\mathbb{Z}G = \pm G$, $T\mathcal{U}\mathbb{Z}G = \pm G$, completing the proof of the theorem. □

3. $\mathcal{U}(\mathbb{Z}G)$ FOR ABELIAN G

We are now in a position to describe the unit group of a commutative integral group ring. See G. Higman [1] and Ayoub and Ayoub [1].

Theorem 3.1. Let G be a finite Abelian group of order n. Then

$$\mathcal{U}\mathbb{Z}G = \pm G \times F,$$

where F is a free Abelian group of rank $\frac{1}{2}(n+1+n_2-2\ell)$. Here, n_d denotes the number of cyclic subgroups of G of order d and $\ell = \Sigma_{d|n}\, n_d$ is the number of cyclic subgroups of G.

Proof: We know from (2.6) that

$$QG = \sum_{d|n}^{\oplus} n_d\, Q(\xi_d) = K,$$

where ξ_d is a primitive d-th root of unity. Then

$$\mathbb{Z}G \subseteq \sum_{d|n} n_d\, \mathbb{Z}[\xi_d] = R$$

and $(\mathcal{U}R : \mathcal{U}\mathbb{Z}G) < \infty$ by (2.9). Hence $\mathcal{U}(\mathbb{Z}G) = T \times F$, where T is finite and F is torsion free of rank equal to the torsion free rank of $\mathcal{U}(R)$. In view of Corollary 1.8 it remains to prove that

F has rank as stated.

By Dirichlet's Unit Theorem (see Hasse [1]), if $d > 2$, $\mathbb{Z}[\xi_d]$ has $\frac{\phi(d)}{2} - 1$ independent units of infinite order. Thus the torsion free rank of $\mathcal{U}(R)$ is

$$= \sum_{d>2} n_d\left(\frac{\phi(d)}{2} - 1\right)$$

$$= \frac{1}{2}\sum_{d>2} n_d\phi(d) - \sum_{d>2} n_d$$

$$= \frac{1}{2}\left(\sum_d n_d\phi(d) - 2\sum_d n_d + n_2 + n_1\right)$$

$$= \frac{1}{2}(n - 2\ell + n_2 + 1). \ \square$$

<u>Lemma 3.2</u>. Let I be an integral domain and let G be torsion free Abelian. Then

$$\mathcal{U}(IG) = \mathcal{U}(I)\cdot G.$$

<u>Proof</u>: Since G can be ordered, the result follows by looking at the largest and the smallest elements $g \in G$ occuring in the support of a unit of IG and its inverse. □

<u>Lemma 3.3</u>. Suppose R is a commutative ring with 0 and 1 as its only idempotents. Suppose $X = \langle x\rangle$ is an infinite cyclic group. Then $\mathcal{U}(RX) = \mathcal{U}(R)\cdot X$ if R contains no nonzero nilpotent elements.

<u>Proof</u>: Suppose $\gamma,\mu \in RX$ such that $\gamma\mu = 1$. We can take $\gamma = \Sigma_0^s a_i x^i$ and $\mu = \Sigma_{-t}^r b_j x^j$, $a_s \neq 0$. We first claim that $\mu = \Sigma_{-s}^0 b_j x^j$, $b_{-s} \neq 0$. Let P be a prime ideal of R which does not contain a_s. Then looking at $\gamma\mu = 1$ modulo PX we conclude by (3.2) that $b_{-s} \neq 0$ and similarly if $b_{-t} \neq 0$ then $a_t \neq 0$. Thus $t = s$. By the same argument if $b_j \neq 0$ for some $j > 0$ then $a_{-j} \neq 0$. We conclude that $\gamma = \Sigma_0^s a_i x^i$ and $\mu = \Sigma_0^s b_{-j} x^{-j}$.

Next we assert that $a_i a_j = 0$ and $a_i b_{-j} = 0$ for $i \neq j$.

Suppose $a_ia_j \neq 0$. Choose a prime ideal P such that $a_ia_j \notin P$. Then $\bar{a}_i \neq 0$ and $\bar{a}_j \neq 0$ in R/P and $\overline{\gamma\mu} = 1$ in the group ring $(R/P)X$. This is a contradiction to (3.2). Thus $a_ia_j = 0$ and similarly $a_ib_{-j} = 0$ and $b_{-i}b_{-j} = 0$ for $i \neq j$.

Now we have the situation

$$\left(\sum_0^s a_ix^i\right)\left(\sum_0^s b_{-j}x^{-j}\right) = 1$$

$$a_ib_{-j} = 0 = a_ia_j = b_{-i}b_{-j} \quad \text{for } i \neq j. \qquad (*)$$

We can suppose $a_sb_{-s} \neq 0$. Then

$$a_0b_0 + a_1b_{-1} + \cdots + a_sb_{-s} = 1.$$

Multiplying by a_s we have

$$a_s^2b_{-s} = a_s \quad \text{and} \quad (a_sb_{-s})^2 = a_sb_{-s}.$$

Thus $a_sb_{-s} = 1$ and from (*) $a_i = 0$ for $i \neq s$. Hence

$$\gamma = a_sx^s \quad \text{and} \quad \mu = b_{-s}x^{-s}. \quad \square$$

Theorem 3.5 (Sehgal [4]). Suppose that G is an arbitrary Abelian group and that R is an integral domain of characteristic 0 satisfying $\mathcal{U}R \cap \{o(g): g \in G\} = \{1\}$. Then

$$\mathcal{U}(RG) = G \cdot \mathcal{U}(RT)$$

where T is the torsion subgroup of G.

Proof: Let $\gamma \in \mathcal{U}(RG)$. We may suppose that G is finitely generated and therefore

$$G = T \times \langle x_1\rangle \times \cdots \times \langle x_s\rangle, \quad |T| < \infty, \quad o(x_i) = \infty.$$

We use induction on s. Since $R(T \times \langle x_1\rangle \times \cdots \times \langle x_{s-1}\rangle)$ has no nontrivial idempotent by (I.2.20) and no nonzero nilpotent element by (2.13), we can apply Lemma 3.3 to conclude that

$$\gamma = \gamma_1 x_s^i, \quad \gamma_1 \in \mathcal{U}(RG_1), \quad G_1 = T \times \langle x_1 \rangle \times \cdots \times \langle x_{s-1} \rangle.$$

Now by induction;

$$\gamma_1 = ug_1, \quad u \in \mathcal{U}(RT), \quad g_1 \in G.$$

Hence,

$$\gamma = ug_1 x_s^i = ug, \quad g \in G, \quad u \in \mathcal{U}(RT). \quad \square$$

Corollary 3.6. If G is a finitely generated Abelian group, say, $G = T \times F$ where T is finite and F is free, then

$$\mathcal{U}(\mathbb{Z}G) = \mathcal{U}(\mathbb{Z}T) \times F.$$

4. TRIVIAL $\mathcal{U}(\mathbb{Z}G)$

Theorem 4.1 (G. Higman [1]). Suppose G is a torsion group. Then $\mathcal{U}(\mathbb{Z}G) = \pm G$ if and only if G is either

(1) an Abelian group with $G^4 = \{1\}$;

or (2) an Abelian group with $G^6 = \{1\}$;

or (3) a Hamiltonian 2-group.

Proof: Suppose that $\mathcal{U}(\mathbb{Z}G) = \pm G$. Then we claim that $\mathbb{Z}G$ has no (nonzero) nilpotent elements. To see this, let $\mu \in \mathbb{Z}G$, $\mu^2 = 0$. Then $(1+\mu)(1-\mu) = 1$ and by hypothesis, $1 + \mu = \pm g$, $1 - \mu = \pm h$, $g,h \in G$. Therefore, $2 = (1+\mu) + (1-\mu) = (\pm g \pm h)$ which implies that $g = h = 1$ and $\mu = 0$. Now it follows by the implication (d) $\Rightarrow$ (e) of (2.14) and (2.18) that G is Abelian or a Hamiltonian 2-group.

First let G be an Abelian group. Then for a subgroup H of order n of G (by the proof of Theorem 3.1) the torsion free part of $\mathcal{U}(\mathbb{Z}H)$ has rank $\rho = \Sigma_{d>2}\, n_d \left(\frac{\phi(d)}{2} - 1\right)$ where n_d is the number of cyclic subgroups of order d of H. It is clear that $\rho = 0$ for every $H \subseteq G$ if and only if G is of the type (1) or (2) of the statement. If G is a Hamiltonian 2-group then we know by (2.5) that $\mathcal{U}(\mathbb{Z}G) = \pm G$ and the proof of the theorem is complete. □

5. PERIODIC NORMAL SUBGROUPS OF THE UNIT GROUP

For any group ring RG, write

$$\mathcal{U}_1(RG) = \{\gamma = \sum \gamma(g)g \in \mathcal{U}(RG) \quad \text{such that} \quad \sum \gamma(g) = 1\}.$$

We call these units of RG, the 1-units. Clearly,

$$\mathcal{U}(RG) = \mathcal{U}(R) \cdot \mathcal{U}_1(RG).$$

The following extension of a part of (2.18) is due to Bovdi [1].

Theorem 5.1. Let I be an integral domain of characteristic 0 such that no rational prime is a unit of I. Then every periodic normal subgroup N of $\mathcal{U}_1(IG)$ is contained in G. Every subgroup of N is normal in G. Further N is Abelian or a Hamiltonian 2-group.

Proof: Let $\alpha = \Sigma\alpha(g)g \in N \lhd \mathcal{U}_1(IG)$. We wish to show that $\alpha \in G$. Without loss of generality, we may assume that α has order p^m where p is a prime. We shall apply the following result which will be proved in (VI.2.1).

(5.2) $\alpha = \Sigma\alpha(g)g \in IG,\ o(\alpha) = p^m \Rightarrow$ there exists $g_0 \in G$ with $o(g_0) = p^m$ and $\alpha(g_0) \neq 0$.

Then consider $\beta = \alpha g_0^{-1} = \Sigma\beta(g)g$ which satisfies $\beta(1) \neq 0$ and

$$\begin{aligned}\beta^s &= (\alpha g_0^{-1})^s \\ &= \alpha g_0^{-1}\alpha g_0^{-1} \cdots \alpha g_0^{-1} \\ &= \alpha\, \alpha^{g_0}\, \alpha^{g_0^2} \cdots \alpha^{g_0^{s-1}}\, g_0^{-s}\ .\end{aligned}$$

By taking $s = p^m$ we get $\beta^s \in N$. Since N is a normal torsion group it follows that β is an element of finite order. It follows by (1.2) that $\beta = 1$ and thus $\alpha = g_0 \in G$.

We shall next show that every cyclic subgroup $\langle h \rangle$ of order n of N is normal in G. Let us write for $g \in G$,

$$\mu = (1+h+\cdots+h^{n-1})g(1-h).$$

Then $\mu^2 = 0$ and $(1+\mu)(1-\mu) = 1$. Therefore,

$$\begin{aligned}(1+\mu)^{-1}h(1+\mu) &= h - \mu h + h\mu - \mu h\mu \\ &= h - \mu h + \mu \\ &= h + (1+h+\cdots+h^{n-1})g(1-h)^2\end{aligned}$$

is an element of N. This implies a relation of the type

$$h^s g = h^t gh \quad \text{for some} \quad s \geq 0 \quad \text{and} \quad t \leq n-1.$$

Hence $ghg^{-1} \in \langle h \rangle$ and $\langle h \rangle$ is normal in G.

Thus N is Abelian or Hamiltonian. Suppose that N is Hamiltonian. We claim that

(5.3) $\gamma = \sum \gamma(g)g \in \mathcal{U}_1(\mathbb{Z}N)$, $\gamma^m = 1$ for some $m \in \mathbb{Z} \Rightarrow \gamma = g \in N$.

Suppose $\gamma(g_1) \neq 0$. Then because N is normal in $\mathcal{U}_1(\mathbb{Z}N)$, γg_1^{-1} is a torsion element of $\mathcal{U}_1(\mathbb{Z}N)$ and is therefore 1 by (1.3). Hence $\gamma = g_1$. We have proved that all torsion units of $\mathbb{Z}N$ are trivial. It follows by (2.18) that N is a 2-group, completing the proof of the theorem. □

Corollary 5.4. $\mathcal{U}(\mathbb{Z}G)$ is periodic if and only if one of the following conditions holds.

(1) G is a Hamiltonian 2-group;

(2) G is an Abelian group with $G^4 = \{1\}$;

(3) G is an Abelian group with $G^6 = \{1\}$.

Proof: (5.1) and (4.1).

We see that only in very special cases $\mathcal{U}(\mathbb{Z}G)$ is periodic. We shall also characterize in Chapter VI groups G for which $\mathcal{U}(\mathbb{Z}G)$ is torsion free. But now we compute $\mathcal{U}(\mathbb{Z}S_3)$, where S_3 is the symmetric group on three elements, in order to get an idea of its size.

6. UNITS OF $\mathbb{Z}S_3$

Let us denote by R_n the full matrix ring of degree n over R. Consider the map

$$\theta\colon QS_3 \to Q \oplus Q \oplus Q_2$$

given by

(6.1)
$$\theta(12) = \left[1,\ -1,\ \begin{bmatrix} 1 & -1 \\ 0 & -1 \end{bmatrix}\right],$$
$$\theta(123) = \left[1,\ 1,\ \begin{bmatrix} 0 & -1 \\ 1 & -1 \end{bmatrix}\right].$$

Keeping in mind that we multiply in S_3 from right to left (e.g. (12)(123) = (23)) we have a homomorphism by linear extension. In fact if $\alpha = (\alpha_1,\cdots,\alpha_6)$ denotes the element

$$\alpha_1 I + \alpha_2(12) + \alpha_3(23) + \alpha_4(13) + \alpha_5(123) + \alpha_6(132)$$

of QS_3 and if $x = (x_1,\cdots,x_6)$ denotes the element

$$\left[x_1,\ x_2,\ \begin{bmatrix} x_3 & x_4 \\ x_5 & x_6 \end{bmatrix}\right]$$

of $Q \oplus Q \oplus Q_2$ and we think of α and x as row vectors then

(6.2) $\quad x = \theta\alpha = \alpha\, A \quad$ where

$$A = \begin{bmatrix} 1 & 1 & 1 & 0 & 0 & 1 \\ 1 & -1 & 1 & -1 & 0 & -1 \\ 1 & -1 & -1 & 0 & -1 & 1 \\ 1 & -1 & 0 & 1 & 1 & 0 \\ 1 & 1 & 0 & -1 & 1 & -1 \\ 1 & 1 & -1 & 1 & -1 & 0 \end{bmatrix}$$

and

$$A^{-1} = \frac{1}{6} \begin{bmatrix} 1 & 1 & 1 & 1 & 1 & 1 \\ 1 & -1 & -1 & -1 & 1 & 1 \\ 2 & 2 & -2 & 0 & -2 & 0 \\ 0 & 0 & -2 & 2 & -2 & 2 \\ 0 & -2 & 0 & 2 & 2 & -2 \\ 2 & -2 & 2 & 0 & 0 & -2 \end{bmatrix} .$$

It follows that θ is an isomorphism. Also, it is clear that

$$\theta\,\mathbb{Z}S_3 \subset \mathbb{Z} \oplus \mathbb{Z} \oplus \mathbb{Z}_2 .$$

Therefore, for $x_i \in \mathbb{Z}$, $\theta^{-1}x \in \mathbb{Z}S_3$ if and only if

$$\begin{aligned}
x_1 + x_2 + 2x_3 \qquad\qquad + 2x_6 &\equiv 0 \pmod 6\\
x_1 - x_2 + 2x_3 \qquad - 2x_5 - 2x_6 &\equiv 0 \pmod 6\\
x_1 - x_2 - 2x_3 - 2x_4 \qquad + 2x_6 &\equiv 0 \pmod 6\\
x_1 - x_2 \qquad + 2x_4 + 2x_5 \qquad &\equiv 0 \pmod 6\\
x_1 + x_2 - 2x_3 - 2x_4 + 2x_5 \qquad &\equiv 0 \pmod 6\\
x_1 + x_2 \qquad + 2x_4 - 2x_5 - 2x_6 &\equiv 0 \pmod 6.
\end{aligned}$$

It is easy to see by row reduction that these congruences are equivalent to

$$\begin{aligned}
&x_1 + x_2 \equiv 0 \pmod 2\\
(6.3)\qquad &x_2 \equiv x_6 - x_5 \pmod 3\\
&x_1 \equiv x_3 + x_5 \equiv x_4 + x_6 \pmod 3.
\end{aligned}$$

If we denote the projection $Q \oplus Q \oplus Q_2 \to Q_2$ by ϕ, we see that

$$\phi\ \theta\ \mathbb{Z}S_3 = \left\{ \begin{bmatrix} x_3 & x_4 \\ x_5 & x_6 \end{bmatrix} : x_3 + x_5 \equiv x_4 + x_6 \pmod 3 \right\}$$

$$= Y \quad \text{(say)}.$$

Suppose $x = (x_1,\cdots,x_6) \in \theta\ \mathbb{Z}S_3$. Then since $x_6 \equiv x_3 + x_5 - x_4 \pmod 3$ it follows that $x_3x_6 - x_4x_5 = \delta$ implies that $(x_3-x_4)(x_3+x_5) \equiv \delta \pmod 3$. It then follows from (6.3) that x^{-1} exists and is in $\theta\ \mathbb{Z}S_3$ if and only if

(6.4) $\qquad x_3x_6 - x_4x_5 = \delta = \pm 1, \quad x_1 = \pm 1, \quad x_2 = \delta x_1.$

The mapping $\phi\theta$ is a ring homomorphism from $\mathbb{Z}S_3$ into Y and so induces a homomorphism $\mathcal{U}(\mathbb{Z}S_3) \to \mathcal{U}(Y)$. In fact this is an isomorphism onto. For let

$$z = \begin{bmatrix} x_3 & x_4 \\ x_5 & x_6 \end{bmatrix} \in \mathcal{U}(Y).$$

Then $\delta = x_3x_6 - x_4x_5 = \pm 1$ and, if x_1, x_2 lying in $\{-1,0,1\}$ are defined by

$$x_2 \equiv x_6 - x_5 \pmod 3$$

$$x_1 \equiv x_3 + x_5 \pmod 3,$$

it follows that neither x_1 nor x_2 is 0 and, in fact, that (6.3) and (6.4) are satisfied. Thus $\alpha = \theta^{-1}x$ is a unit in $\mathbb{Z}S_3$ with $\phi\theta\alpha = z$. Further it is a consequence of (6.3) and (6.4) that $\phi\theta$ is one-to-one. We have proved

Theorem 6.5 (Hughes and Pearson [1])

$$\mathcal{U}(\mathbb{Z}S_3) \simeq \left\{ \begin{bmatrix} a & b \\ c & d \end{bmatrix} \in \mathcal{U}(\mathbb{Z}_2) \;\middle|\; a + c \equiv b + d \pmod 3 \right\}.$$

CHAPTER III

THE ISOMORPHISM PROBLEM

It is natural to ask as to what extent RG reflects the properties of G. Can we have RG isomorphic to RH as an R-algebra (written $RG \simeq_R RH$) with G not isomorphic to H? A moment's reflection tells us that if G and H are finite Abelian groups of the same order then

$$\mathbb{C}G \underset{\mathbb{C}}{\simeq} \mathbb{C}H \underset{\mathbb{C}}{\simeq} \mathbb{C} \oplus \cdots \oplus \mathbb{C}.$$

The reason for this occurence is that the ring of coefficients $\mathbb{C}$ is too large. Also, the rational group algebras of the two non-commutative groups of order p^3 are isomorphic. The proper rings R to consider seem to be $\mathbb{Z}$ and for p-groups $\mathbb{Z}/p\mathbb{Z}$ and the p-adic integers $\mathbb{Z}_p$. We state the well known

Isomorphism Problem. Is it true that for a suitable commutative ring R, the isomorphism of R-algebras,

$$RG \underset{R}{\simeq} RH \quad \text{implies} \quad G \simeq H?$$

Remark. Suppose that $RG \overset{\theta}{\underset{R}{\simeq}} RH$. Define $\theta^*: RG \to RH$ by

$$\theta^*\left(\sum a_g\, g\right) = \sum a_g \left(c\theta(g)\right)^{-1}\theta(g) \quad \text{where} \quad c\left(\sum b_h\, h\right) = \sum b_h$$

is the augmentation of $\Sigma b_h h$. Then θ^* is an R-isomorphism and preserves augmentation i.e. $c(\theta^*(g)) = 1$ for all $g \in G$. Such an isomorphism is called normalized. We shall assume without loss of generality, that all isomorphisms $RG \underset{R}{\simeq} RH$ in this chapter are normalized.

We first give a brief account of needed facts regarding dimension subgroups.

1. DIMENSION SUBGROUPS

Let R be any commutative ring with identity, G any group and i any natural number. Then by the i-th dimension subgroup of G over R, written $D_{i,R}(G)$, we understand

$$D_{i,R}(G) = \{g \in G\colon g-1 \in \Delta_R^i(G)\} = G \cap \left(1 + \Delta_R^i(G)\right),$$

where $\Delta_R^i(G)$ is the i-th power of $\Delta_R^i(G)$.
It follows by using the identities

$$xy - 1 = (x-1)(y-1) + (x-1) + (y-1)$$
$$xgx^{-1} - 1 = x(g-1)x^{-1}$$

that $D_{i,R}(G)$ is a normal subgroup of G, giving us a normal series

$$G = D_{1,R}(G) \supseteq D_{2,R}(G) \supseteq \cdots \supseteq D_{n,R}(G) \supseteq \cdots.$$

Moreover, $\gamma_i(G) \subseteq D_{i,R}(G)$ because $\{D_{i,R}(G)\}$ is a central series as seen below.

<u>Lemma 1.1</u>. (i) $\left(D_{i,R}(G), D_{j,R}(G)\right) \subseteq D_{i+j,R}(G)$ for all $i,j \geq 1$.

(ii) If the characteristic of R is a prime p then

$$\left(D_{i,R}(G)\right)^p \subseteq D_{ip,R}(G).$$

<u>Proof:</u> (i) follows from the identity

$$x^{-1}y^{-1}xy-1 = x^{-1}y^{-1}(xy-yx) = x^{-1}y^{-1}[(x-1)(y-1)-(y-1)(x-1)].$$

(ii) follows by observing that if R has characteristic p then for $g \in G$,

$$(g^p-1) = (g-1)^p. \quad \square$$

We shall prove that in order to compute $D_{i,R}(G)$ it is enough to confine our attention to $D_{i,\mathbb{Z}}(G)$ and $D_{i,\mathbb{Z}/n\mathbb{Z}}(G)$. To state and prove this result let us fix notation for the rest of this section.

R is a fixed commutative ring with identity. We denote by $\sigma(R)$ the set of all rational primes p satisfying $p^m R = p^{m+1} R$ for some non-negative integer m. Write $\Pi(R) = \sigma(R) \cap \mathcal{U}R$, where $\mathcal{U}R$ is the unit group of R i.e. $\Pi(R)$ is the set of those rational primes p such that there exists an $r \in R$ with $pr = 1$. For $p \in \sigma(R)$, let us denote by $e = e(p)$ the smallest non-negative integer m satisfying $p^m R = p^{m+1} R$. We need a few lemmas.

<u>Lemma 1.2.</u> Let $I = \{r \in R\colon zr = 0$ for some $z \in \mathbb{Z}$, $z \neq 0\}$. Then I is an ideal and $\sigma(R/I) = \sigma(R)$.

<u>Proof</u>: Clearly I is an ideal of R and $\sigma(R) \subseteq \sigma(R/I)$. Conversely, let us suppose that $p \in \sigma(R/I)$. Then $p^m(R/I) = p^{m+1}(R/I)$ for some m. We have, therefore,

$$p^m = p^{m+1}r + i, \quad \text{for some } r \in R, \quad i \in I.$$

Then for a suitable $z \in \mathbb{Z}$, $p^m z = p^{m+1} rz$. Let us write,

$$p^m z = p^i k, \quad (p,k) = 1, \quad i \geq m \geq 0.$$

Choose integers a and b such that $ak + bp = 1$. Then

$$p^{i+1}akr = p^{m+1}rza = p^m za = p^i ka = p^i(1-bp).$$

Consequently, $p^i \in p^{i+1}R$ and $p \in \sigma(R)$, completing the proof of the lemma.

Lemma 1.3. Suppose that for a prime p, $p^m R = p^{m+1} R$ for some non-negative integer m. Let $J = \{r \in R: p^m r = 0\}$. Then J is an ideal of R such that

$$R \simeq R/p^m R \oplus R/J, \quad \sigma(R) = \sigma(R/J) \text{ and } p \in \Pi(R/J).$$

Proof: We first assert that $R = p^m R \oplus J$. Given $r \in R$, $p^m r = p^{2m} s$ for a suitable $s \in R$. Therefore $p^m(r - p^m s) = 0$. Thus $r = p^m s + (r - p^m s)$ and $R \subseteq p^m R + J$. Now,

$$a \in p^m R \cap J \Rightarrow a = p^m b, \quad p^m(p^m b) = 0, \quad b \in R.$$

But $a = p^m b = p^{2m} tb$ for a suitable $t \in R$, which implies that $a = 0$. We have thus proved that $R = p^m R \oplus J$. Let us define a homomorphism,

$$\theta: R \to R/p^m R \oplus R/J$$

by, $\theta(r) = (r + p^m R, r + J)$ for $r \in R$. Then

$$\theta(R) = \theta(p^m R + J) = R/p^m R \oplus R/J$$

and θ is 1-1 as $R = p^m R \oplus J$. Clearly, $\sigma(R) \subseteq \sigma(R/J)$. Also since $p^m R = p^{m+1} R$, $p^m = p^{m+1} r$ for some $r \in R$. Then $p^m(pr - 1) = 0$ and $p \in \Pi(R/J)$. It remains to prove that

$$q \neq p, \quad q \in \sigma(R/J) \Rightarrow q \in \sigma(R).$$

Let $q^t = q^{t+1} r + j$, $r \in R$, $j \in J$. Choose $a, b \in \mathbb{Z}$ such that $p^m a + q^{t+1} b = 1$. Then,

$$q^t = (p^m a + q^{t+1} b) q^t = (p^m a + q^{t+1} b)(q^{t+1} r + j) \in q^{t+1} R$$

as $p^m j = 0$. Hence $q \in \sigma(R)$. □

Let us denote by $A - B$, the complement of a set B in a set $A \supseteq B$.

Lemma 1.4. R Noetherian $\Rightarrow |\sigma(R) - \Pi(R)| < \infty$.

Proof: Suppose that there exists an infinite sequence $p_1, p_2, \cdots, p_m, \cdots$ of distinct primes in $\sigma(R) - \Pi(R)$. Write

$$p_i^{n_i}R = p_i^{n_i+1}R, \quad t_m = \prod_{i=1}^{m} p_i^{n_i}, \quad J_m = \{r \in R: t_m r = 0\}.$$

We assert that the ascending chain

$$J_1 \subseteq J_2 \subseteq \cdots \subseteq J_m \subseteq \cdots$$

of ideals of R does not become stationary after a finite number of steps. Let $p_m^{n_m} = p_m^{n_m+1} v_m$ for a suitable $v_m \in R$. Then $1 - p_m v_m \in J_m$ but $1 - p_m v_m \notin J_{m-1}$. Otherwise $(1-p_m v_m)t_{m-1} = 0$ which would imply, due to the relative primeness of t_{m-1} and p_m, that $1 - p_m v_m = 0$ which is not so as $p_m \notin \Pi(R)$. Thus we have proved that if $\sigma(R) - \Pi(R)$ is infinite then R is not Noetherian, proving the lemma.

Lemma 1.5. If R is a direct sum of two rings, $R = R_1 \oplus R_2$ then

(i) $RG \simeq R_1G \oplus R_2G$,

(ii) $g \in D_{n,R}(G) \Longleftrightarrow g \in D_{n,R_1}(G)$ and $g \in D_{n,R_2}(G)$.

Proof: Define $\sigma: RG \to R_1G \oplus R_2G$ by

$$\sigma\left(\sum_i (r_i, s_i)g_i\right) = \left(\sum_i r_i g_i, \sum_i s_i g_i\right).$$

It is easy to check that σ is an isomorphism.

(ii) Let us identify R_1 and R_2 with $(R_1,0)$ and $(0,R_2)$ respectively. Then $R_1G = (R_1,0)G$ and $R_2G = (0,R_2)G$. For $x \in G$, the element $x - 1$ of RG can be written as

$$(1,1)x - (1,1) = [(1,0)x - (1,0)] + [(0,1)x - (0,1)] = \alpha + \beta$$

with $\alpha \in \Delta_{R_1}(G)$ and $\beta \in \Delta_{R_2}(G)$. Thus if $g \in D_{n,R}(G)$, we have by taking homomorphic images that $g \in D_{n,R_1}(G)$ and $g \in D_{n,R_2}(G)$. Conversely, let $g \in D_{n,R_1}(G)$ and $g \in D_{n,R_2}(G)$. Then

$$(1,0)g - (1,0) \in \Delta^n_{R_1}(G) \quad \text{and} \quad (0,1)g - (0,1) \in \Delta^n_{R_2}(G).$$

But $\Delta^n_{R_1}(G) \subseteq \Delta^n_R(G)$ and $\Delta^n_{R_2}(G) \subseteq \Delta^n_R(G)$ as the augmentation in RG of $[(1,0)x - (1,0)]$ and $[(0,1)x - (0,1)]$ is zero for each $x \in G$. Thus

$$g-1 = (1,1)g - (1,1) = (1,0)g - (1,0) + (0,1)g - (0,1) \in \Delta^n_R(G)$$

and $g \in D_{n,R}(G)$. □

Recall that for $g \in G$, we understand by $g - 1$ the element $1_R g - 1_R$ of the group ring RG where 1_R is the unity element of R. We shall have to deal with elements $1_R g - 1_R \in RG$ and $1_S g - 1_S \in SG$ for two different rings R and S. However, we shall denote them both by $(g-1)$ as it will be clear from the context as to which group ring this element belongs to.

Lemma 1.6 (Sandling [3]). If R has characteristic $m > 0$ then, for $g \in G$,

$$g-1 \in \Delta^n_R(G) \Rightarrow g-1 \in \Delta^n_{\mathbb{Z}/m\mathbb{Z}}(G).$$

Proof: Suppose that $(g-1) \notin \Delta^n_{\mathbb{Z}/m\mathbb{Z}}(G)$. Then $\overline{g-1} = g-1 + \Delta^n_{\mathbb{Z}/m\mathbb{Z}}(G)$ is a nonzero element of the additive group $\mathbb{Z}/m\mathbb{Z}(G)/\Delta^n_{\mathbb{Z}/m\mathbb{Z}}(G)$. Therefore, there exists a nonzero homomorphism from $\langle\overline{g-1}\rangle$ to $T = Q/\mathbb{Z}$, the additive group of the rationals mod 1. Since T is divisible, this homomorphism can be extended to

$$\lambda\colon \mathbb{Z}/m\mathbb{Z}(G)/\Delta^n_{\mathbb{Z}/m\mathbb{Z}}(G) \to T.$$

Clearly, the image of λ is contained in a cyclic group of order m and we may regard

$$\lambda: \mathbb{Z}/m\mathbb{Z}(G)/\Delta^n_{\mathbb{Z}/m\mathbb{Z}}(G) \to \mathbb{Z}/m\mathbb{Z}$$

with $\lambda(\overline{g-1}) \neq 0$. Define a map $\mu: G \to \mathbb{Z}/m\mathbb{Z}$ by $\mu(x) = \lambda(\overline{x})$. Then we have,

$$\tilde{\mu}: G \xrightarrow{\mu} \mathbb{Z}/m\mathbb{Z} \xrightarrow{i} R$$

where the last map is the natural injection. Extending $\tilde{\mu}$ R-linearly we have

$$\tilde{\mu}: RG \to R$$

with

$$\tilde{\mu}(g-1) = \tilde{\mu}(g) - \tilde{\mu}(1) = \lambda(\overline{g}) - \lambda(\overline{1}) = \lambda(\overline{g-1}) \neq 0.$$

But $\tilde{\mu}(\Delta^n_R(G)) = 0$ as for $\Sigma a_h\ h = (g_1-1)\cdots(g_n-1) \in RG$, $g_i \in G$, $\tilde{\mu}(\Sigma a_h\ h) = \Sigma a_h\ \tilde{\mu}(h) = \Sigma a_h\ \lambda(\overline{h}) = \lambda(\Sigma a_h\ \overline{h}) = \lambda(\overline{\delta})$, $\delta = (g_1-1)\cdots(g_n-1)$. Therefore, $\tilde{\mu}(g-1) = 0$ which is a contradiction proving the result.

<u>Lemma 1.7</u>. Suppose that R has characteristic 0 and that all rational primes are in $\Pi(R)$. Then

$$g-1 \in \Delta^n_R(G) \Rightarrow k(g-1) \in \Delta^n_{\mathbb{Z}}(G)\ ,$$

for a suitable natural number k.

<u>Proof</u>: Suppose that $k(g-1) \notin \Delta^n_{\mathbb{Z}}(G)$ for any natural number k. Then $\overline{g-1} = g-1 + \Delta^n_{\mathbb{Z}}(G)$ is an element of infinite order of the additive group $\mathbb{Z}G/\Delta^n_{\mathbb{Z}}(G)$. Thus there exists a nonzero additive homomorphism of $\langle\overline{g-1}\rangle$ into Q which we can extend to

$$\lambda: \mathbb{Z}G/\Delta^n_{\mathbb{Z}}G \to Q.$$

Define $\mu: G \to Q$ by $\mu(x) = \lambda(\bar{x})$. Then we have

$$\tilde{\mu}: G \to Q \to R$$

the last map being the natural injection. Extending $\tilde{\mu}$ R-linearly we have

$$\tilde{\mu}: RG \to R$$

with $\tilde{\mu}(g-1) = \tilde{\mu}(g) - \tilde{\mu}(1) = \lambda(\bar{g}) - \lambda(\bar{1}) = \lambda(\overline{g-1}) \neq 0$. On the other hand, since it can be easily checked that $\tilde{\mu}(\Delta_R^n(G)) = 0$, $\tilde{\mu}(g-1) = 0$. This is a contradiction proving the lemma. □

Now we state the main theorem of this section. This result was proved for groups G satisfying $\gamma_n(G) = D_{n,\mathbb{Z}}(G)$ by Sandling [3] and in general by Parmenter [2]. A unified proof for this and an analogous result for Lie dimension subgroups is contained in a paper of Parmenter, Passi and Sehgal [1].

In addition to the notation set out at the beginning of this section we need the

Definition 1.8. For a normal subgroup N of G and β a set of primes we understand by $T_\beta(G \bmod N)$ the β-torsion of $G \bmod N$ i.e.

$$T_\beta(G \bmod N) = \{g \in G\colon g^k \in N \text{ for some } \beta\text{-number } k\}.$$

A β-number is a natural number divisible only by primes in β. If β is the set of all primes we denote $T_\beta(G \bmod N)$ by $\sqrt{N}$.

Theorem 1.9 (Parmenter-Sandling). Let R be a commutative ring with identity and G any group.

(i) If the characteristic of R is $m > 0$,

$$D_{n,R}(G) = D_{n,\mathbb{Z}/m\mathbb{Z}}(G).$$

(ii) If the characteristic of R is 0,

$$D_{n,R}(G) = \prod_{p \in \sigma(R)} T_p\big(G \bmod D_{n,\mathbb{Z}}(G)\big) \cap D_{n,\mathbb{Z}/p^e\mathbb{Z}}(G).$$

(If $\sigma(R)$ is empty the right hand side is to be interpreted as $D_{n,\mathbb{Z}}(G)$).

The following is a crucial lemma.

Lemma 1.10. Suppose that R has characteristic 0. Then for $g \in G$ and n any natural number,

$$g-1 \in \Delta_R^n(G) \Rightarrow k(g-1) \in \Delta_{\mathbb{Z}}^n(G), \text{ for a suitable } \sigma(R)\text{-number } k.$$

(If $\sigma(R)$ is empty, k is to be interpreted as 1).

Proof: We may assume that R is Noetherian, in particular, generated by all coefficients in an equation of the form

$$g-1 = \sum a(g_1-1)(g_2-1) \cdots (g_n-1), \quad a \in R, \quad g_i \in G.$$

Then by Lemma 1.4, $|\sigma(R) - \Pi(R)| < \infty$. We shall prove the lemma by induction on $|\sigma(R) - \Pi(R)|$. First, suppose that $\sigma(R) = \Pi(R)$. If $\Pi(R)$ is the set of all primes the result follows by Lemma 1.7. We may, therefore, assume that there exists a prime $q \notin \Pi(R)$. Let $I = \{r \in R: mr = 0 \text{ for some } m \in \mathbb{Z}, m \neq 0\}$. Then $(R/I,+)$, the additive group of R/I is torsion free and thus by Lemma 1.2,

$$\Pi(R/I) = \sigma(R/I) = \sigma(R) = \Pi(R).$$

Replacing R by R/I we can further suppose that $(R,+)$ is torsion free. Suppose that for no $\sigma(R)$-number k, $k(g-1) \in \Delta_{\mathbb{Z}}^n(G)$. Then $\overline{g-1} = g-1 + \Delta_{\mathbb{Z}}^n(G)$ is an element of the group $\mathbb{Z}G/\Delta_{\mathbb{Z}}^n(G)$ whose order is infinite or has a prime divisor $p \notin \Pi(R)$. In any case, we have a nonzero homomorphism of $\langle\overline{g-1}\rangle$ to $Z(p^\infty)$, the Sylow p-subgroup of the additive group $Q/\mathbb{Z}$, of rationals mod 1 where $p \notin \Pi(R)$. Since $Z(p^\infty)$ is divisible this homomorphism can be extended to

$$f: \mathbb{Z}G/\Delta_{\mathbb{Z}}^n(G) \to Z(p^\infty).$$

Let us denote by R_p the ring of fractions $\{r/p^a,\ r \in R,\ a \geq 0\}$. Then we have natural imbeddings $R \to R_p$ and $Z(p^\infty) \to R_p/R$. Let us set for $\gamma \in \mathbb{Z}G$, $\alpha(\gamma) = f(\overline{\gamma})$ then we have

$$\alpha\colon \mathbb{Z}G \to Z(p^\infty), \quad \alpha\big(\Delta^n(G)\big) = 0.$$

Let β be the restriction of α to G, followed by the natural injection $Z(p^\infty) \to R_p/R$. We have

$$G \to Z(p^\infty) \to R_p/R,$$

the R-linear extension of which gives

$$\tilde{\beta}\colon RG \to R_p/R.$$

Then $\tilde{\beta}(g-1) = f(\overline{g-1}) \neq 0$. On the other hand, $\tilde{\beta}(\Delta_R^n) = 0$ and thus $\tilde{\beta}(g-1) = 0$ which is a contradiction proving the lemma when $\sigma(R) = \Pi(R)$.

Now let us suppose that $|\sigma(R) - \Pi(R)| \geq 1$. Pick a prime $p \in \sigma(R)$, $p \notin \Pi(R)$. Then by Lemma 1.3,

$$R = R/p^eR \oplus R/J$$

where $J = \{r \in R\colon p^e r = 0\}$. Since also $\sigma(R/J) = \sigma(R)$ and $p \in \Pi(R/J)$, we have $|\sigma(R/J) - \Pi(R/J)| < |\sigma(R) - \Pi(R)|$. By taking a homomorphic image by Lemma 1.5, $g-1 \in \Delta^n_{R/J}(G)$. Therefore, by induction,

$$k(g-1) \in \Delta^n_{\mathbb{Z}}G$$

for a suitable $\sigma(R/J)$-number k. The proof is complete as $\sigma(R/J) = \sigma(R)$. □

<u>Lemma 1.11</u>. Let $g \in G$, $p \in \sigma(R)$, $n, \nu \geq 0$,

$$g^{p^\nu} - 1 \in \Delta^n_{\mathbb{Z}}(G), \quad g-1 \in \Delta^n_{\mathbb{Z}/p^e\mathbb{Z}}(G) \Rightarrow g-1 \in \Delta^n_R(G).$$

<u>Proof:</u> We have

$$g^{p^\nu}-1 = (g-1+1)^{p^\nu} - 1 = p^\nu(g-1) + \binom{p^\nu}{2}(g-1)^2 + \cdots + (g-1)^{p^\nu}.$$

Suppose first that $p \in \Pi(R)$ i.e. $pr = 1$ for some $r \in R$ then

$$g-1 = r^\nu(g^{p^\nu}-1) - r^\nu\binom{p^\nu}{2}(g-1)^2 - \cdots - r^\nu(g-1)^{p^\nu}.$$

Consequently,

$$g^{p^\nu}-1 \in \Delta^n_{\mathbb{Z}}(G) \Rightarrow g^{p^\nu}-1 \in \Delta^n_R(G) \Rightarrow g-1 \in (g-1)^2RG + \Delta^n_R(G) \Rightarrow g-1 \in \Delta^n_R(G).$$

Now suppose that p is not invertible in R. Then since $p \in \sigma(R)$, by Lemmas 1.3 and 1.5, we have

$$R = R/p^eR \oplus R/J, \quad RG \simeq R/p^eRG \oplus R/JG$$

where $J = \{r \in R\colon p^e r = 0\}$. Therefore,

$$g^{p^\nu}-1 \in \Delta^n_{\mathbb{Z}}(G) \Rightarrow g^{p^\nu}-1 \in \Delta^n_R(G) \Rightarrow g^{p^\nu}-1 \in \Delta^n_{R/J}(G) \Rightarrow g-1 \in \Delta^n_{R/J}(G),$$

the last implication follows as above as $p \in \Pi(R/J)$. Since also, $g-1 \in \Delta^n_{\mathbb{Z}/p^e\mathbb{Z}}(G)$, it follows by Lemma 1.5 that $g-1 \in \Delta^n_R(G)$. □

1.12. Proof of Theorem 1.9

(i) Characteristic of $R = m > 0$. Lemma 1.6 implies that $D_{n,R}(G) \subseteq D_{n,\mathbb{Z}/m\mathbb{Z}}(G)$ and the reverse inclusion is trivial.

(ii) Characteristic of $R = 0$. If $n = 1$, there is nothing to prove. Let $n > 1$. Suppose that $g \in D_{n,R}(G)$. Then by Lemma 1.10, $k_1(g-1) \in \Delta^n_{\mathbb{Z}}(G)$ for a suitable $\sigma(R)$-number k_1. Therefore, we can conclude from the equation

$$g^{k_1}-1 = k_1(g-1) + \binom{k_1}{2}(g-1)^2 + \cdots + (g-1)^{k_1},$$

that $g^{k_1}-1 \in \Delta^{\min(n,2)}_{\mathbb{Z}}(G)$. Now we can apply the same argument to $g^{k_1} \in D_{n,R}(G)$ to deduce that there exists a $\sigma(R)$-number k_2 such that

$$k_2(g^{k_1}-1) \in \Delta^n_{\mathbb{Z}}(G), \quad g^{k_1k_2}-1 \in \Delta^{\min(n,4)}_{\mathbb{Z}}(G).$$

Repeating this process a finite number of times we obtain a $\sigma(R)$-number $\ell = k_1 k_2 \cdots k_t$ with $g^{\ell}-1 \in \Delta^n_{\mathbb{Z}}(G)$. Hence,

$$g \in T_{\sigma(R)}\big(G \bmod D_{n,\mathbb{Z}}(G)\big) = \prod_{p\in\sigma(R)} T_p\big(G \bmod D_{n,\mathbb{Z}}(G)\big),$$

since $D_{n,\mathbb{Z}}(G) \supseteq \gamma_n(G)$, the n-th term of the lower central series of G, and $G/D_{n,\mathbb{Z}}(G)$ is nilpotent. (We remark, that if $\sigma(R)$ is empty then this reduces to saying that $g \in D_{n,\mathbb{Z}}(G)$ and since $D_{n,\mathbb{Z}}(G) \subseteq D_{n,R}(G)$, the theorem is proved in this case.)

Conversely, let us suppose that we have an element

$$g \in T_p\big(G \bmod D_{n,\mathbb{Z}}(G)\big) \cap D_{n,\mathbb{Z}/p^e\mathbb{Z}}(G).$$

Then there exists a non-negative integer ν such that

$$g^{p^\nu}-1 \in \Delta^n_{\mathbb{Z}}(G), \quad g-1 \in \Delta^n_{\mathbb{Z}/p^e\mathbb{Z}}(G)$$

and hence by Lemma 1.11, $g \in D_{n,R}(G)$. This completes the proof of the theorem. □

The next proposition gives a further reduction for the nonzero characteristic case of Theorem 1.9.

Proposition 1.13. For an arbitrary group G, we have

$$D_{n,\mathbb{Z}/m\mathbb{Z}}(G) = \bigcap_i D_{n,\mathbb{Z}/p_i^{e_i}\mathbb{Z}}(G)$$

where $m = p_1^{e_1} p_2^{e_2} \cdots p_s^{e_s}$ is the prime power decomposition of the natural number m.

Proof: Clearly, $g \in D_{n,\mathbb{Z}/m\mathbb{Z}}(G)$ implies that $g \in D_{n,\mathbb{Z}/p_i^{e_i}\mathbb{Z}}(G)$ for each i. The reverse inclusion is a direct consequence of Lemma 1.5 as $\mathbb{Z}/m\mathbb{Z} = \Sigma^{\oplus}\mathbb{Z}/p_i^{e_i}\mathbb{Z}$. □

Theorem 1.9, Proposition 1.13 and the next result evaluate completely the second dimension subgroup, $D_{2,R}(G)$.

Proposition 1.14. For an arbitrary group G we have

(i) $D_{2,\mathbb{Z}}(G) = G'$, the derived subgroup of G;

(ii) $D_{2,\mathbb{Z}/p^s\mathbb{Z}}(G) = G'G^{p^s}$, where G^{p^s} denotes the subgroup of G generated by all g^{p^s}, $g \in G$.

Proof: (i) In view of Lemma 1.1(i), $G' = (G,G) \subseteq D_{2,\mathbb{Z}}(G)$. Conversely, suppose that we have a $g \in G$, $g \notin G'$. We shall prove that $g-1 \notin \Delta^2_{\mathbb{Z}}(G)$. Since $\bar{g} = gG'$ is a non-identity element of G/G', we have a nonzero homomorphism $\langle \bar{g} \rangle \to T = \mathbb{Q}/\mathbb{Z}$. Since T is divisible we may extend this to a homomorphism

$$\lambda\colon G/G' \to T, \quad \lambda(\bar{g}) \neq 0.$$

Setting $\mu(x) = \lambda(\bar{x})$ for $x \in G$ and extending this linearly we have an additive homomorphism $\mu\colon \mathbb{Z}G \to T$. Since for $x,y \in G$, $\mu\big((x-1)(y-1)\big) = \mu(xy-x-y+1) = \lambda(\overline{xy}) - \lambda(\bar{x}) - \lambda(\bar{y}) - \lambda(\bar{1}) = 0$, we have that $\mu\big(\Delta^2_{\mathbb{Z}}(G)\big) = 0$. Now, $\mu(g-1) = \lambda(\bar{g}) - \lambda(\bar{1}) = \lambda(\bar{g}) \neq 0$ and therefore $g-1 \notin \Delta^2_{\mathbb{Z}}(G)$. This proves (i).

(ii) $G'G^{p^s} \subseteq D_{2,\mathbb{Z}/p^s\mathbb{Z}}(G)$, in view of the fact that $G' \subseteq D_{2,R}(G)$ for any R and the identities

$$g^{p^s}-1 = (g-1)^{p^s} + \binom{p^s}{1}(g-1)^{p^s-1} + \cdots + \binom{p^s}{p^s-2}(g-1)^2 + p^s(g-1)$$

$$xy-1 = (x-1) + (y-1) + (x-1)(y-1).$$

It remains to prove that if $g \in G$, $g \notin G'G^{p^s}$ then $g-1 \notin \Delta^2_{\mathbb{Z}/p^s\mathbb{Z}}(G)$. Since $g \notin G'G^{p^s}$, we have a nonzero homomorphism $\langle \bar{g} \rangle \to Z(p^\infty)$ where $\bar{g} = gG'G^{p^s}$ is an element of $G/G'G^{p^s}$. Extend this to a homomorphism

$$\lambda\colon G/G'G^{p^s} \to Z(p^\infty).$$

Clearly the image of λ is contained in $\mathbb{Z}/p^s\mathbb{Z}$. So we may regard

$$\lambda\colon G/G'G^{p^s} \to \mathbb{Z}/p^s\mathbb{Z}.$$

Setting for $x \in G$, $\mu(x) = \lambda(\overline{x})$ and extending linearly we have an additive homomorphism

$$\mu: (\mathbb{Z}/p^s\mathbb{Z})G \to \mathbb{Z}/p^s\mathbb{Z}$$

with $\mu(g-1) = \lambda(\overline{g}) - \lambda(\overline{1}) = \lambda(\overline{g}) \neq 0$. Since $\mu\big(\Delta^2_{\mathbb{Z}/p^s\mathbb{Z}}(G)\big) = 0$, it follows that $g-1 \notin \Delta^2_{\mathbb{Z}/p^s\mathbb{Z}}(G)$. This completes the proof of the proposition. □

<u>Proposition 1.15</u>. For $R = \mathbb{Z}/m\mathbb{Z}$, $m \neq 1$ (notice that $\mathbb{Z}/0\mathbb{Z} = \mathbb{Z}$) and N a normal subgroup of an arbitrary group G we have the following isomorphisms of commutative groups:

(i) $\Delta_R(G)/\Delta^2_R(G) \simeq G/D_{2,R}(G)$,

(ii) $\Delta_R(G,N)/\Delta_R(G)\Delta_R(G,N) \simeq N/D_{2,R}(N)$.

<u>Proof</u>: In view of the identity

$$gh-1 = (g-1) + (h-1) + (g-1)(h-1) \tag{1.16}$$

and its consequence

$$0 = (g-1) + (g^{-1}-1) + (g-1)(g^{-1}-1) \tag{1.17}$$

it follows that for $a_i \in R$, $g_i \in G$,

$$\sum a_i(g_i-1) \equiv (\Pi g_i^{a_i}) - 1 \bmod \Delta^2_R(G). \tag{1.18}$$

Thus the map

$$G \ni g \to g-1 + \Delta^2_R(G) \in \Delta_R(G)/\Delta^2_R(G)$$

is an epimorphism whose kernel is $G \cap \big(1 + \Delta^2_R(G)\big) = D_{2,R}(G)$. This proves (i). In order to prove (ii) let us consider the map

$$N \ni n \xrightarrow{\lambda} n-1 + \Delta_R(G)\Delta_R(G,N) \in \Delta_R(G,N)/\Delta_R(G)\Delta_R(G,N).$$

λ is a homomorphism because of (1.16). Also λ is onto because of the equation

$$g(n-1) = n-1 + (g-1)(n-1), \quad g \in G, \quad n \in N,$$

in view of (1.16) and (1.17). To complete the proof it suffices to prove

(1.19) $\quad n \in N, \quad (n-1) \in \Delta_R(G)\Delta_R(G,N) \Rightarrow n \in D_{2,R}(N).$

Accordingly, let us write

(1.20) $\quad n-1 = \sum a_i(g_i-1)(n_i-1), \quad a_i \in R, \quad g_i \in G, \quad n_i \in N.$

Let us choose a transversal X containing 1, for N in G. Then setting for $g \in G$, $g^\theta = n$ if $g = xn$, $x \in X$, $n \in N$ and extending R-linearly we have a R-linear map $\theta\colon RG \to RN$. Applying θ to both sides of (1.20) we obtain

$$\begin{aligned} n-1 &= \sum a_i\big((x_{j_i} n_{j_i} - 1)(n_i-1)\big)^\theta \text{ where } g_i = x_{j_i} n_{j_i},\ x_{j_i} \in X,\ n_{j_i} \in N, \\ &= \sum a_i(x_{j_i} n_{j_i} n_i - x_{j_i} n_{j_i} - n_i + 1)^\theta \\ &= \sum a_i(n_{j_i} n_i - n_{j_i} - n_i + 1) \\ &= \sum a_i(n_{j_i} - 1)(n_i - 1) \in \Delta_R^2(N). \end{aligned}$$

Thus $n \in D_{2,R}(N)$, proving (1.19). □

It follows from Theorem 1.9 that the dimension subgroups $D_{n,K}(G)$ of G over a field K depend only on the characteristic of K. In other words,

$$D_{n,K}(G) = \begin{cases} D_{n,Q}(G) & \text{if } K \text{ has characteristic } 0, \\ D_{n,\mathbb{Z}/p\mathbb{Z}}(G) & \text{if } K \text{ has characteristic } p > 0. \end{cases}$$

Also, $D_{n,K}(G)$ can be completely described in terms of the structure of G. We need a couple of definitions. Let p be a fixed prime. We define the Brauer-Jennings-Zassenhaus M-series $\{M_{n,p}(G)\}_{n\geq 1}$ of a group G inductively by $M_{1,p}(G) = G$ and for $n \geq 2$

$$M_{n,p}(G) = (G, M_{n-1,p}(G))M_i^p(G)$$

where i is the smallest integer satisfying $ip \geq n$. This is then the minimal central series of G with the property:

$$x \in M_{n,p}(G) \Rightarrow x^p \in M_{np,p}(G) \quad \text{for all} \quad n \geq 1. \qquad (*)$$

The Lazard series $\{L_{n,p}(G)\}_{n\geq 1}$, of G is given by

$$L_{n,p}(G) = \prod_{ip^j \geq n} \gamma_i(G)^{p^j}.$$

Then it is easy to see the

Proposition 1.21. For any group G we have

(i) $D_{n,Q}(G) \supseteq \sqrt{\gamma_n(G)}$,
(ii) $D_{n,\mathbb{Z}/p\mathbb{Z}}(G) \supseteq M_{n,p}(G) \supseteq L_{n,p}(G)$.

Proof: (i) Suppose that $g \in \sqrt{\gamma_n(G)}$ i.e. $g^m \in \gamma_n(G)$ for some $m > 0$. Thus $g^m - 1 \in \Delta_Q^n(G)$. It follows from the equation

$$g^m - 1 = m(g-1) + \binom{m}{2}(g-1)^2 + \cdots + \binom{m}{m-1}(g-1)^{m-1}$$

that $g-1 \in \Delta_Q^n(G) + (g-1)^2 QG$. Hence $g-1 \in \Delta_Q^n(G)$ and $g \in D_{n,Q}(G)$.

(ii) Let $g \in \gamma_i(G)$, $ip^j \geq n$ then $g^{p^j} \in M_{ip^j,p}(G) \subseteq M_{n,p}(G)$. Thus

$$L_{n,p}(G) \subseteq M_{n,p}(G).$$

The containment $M_{n,p}(G) \subseteq D_{n,\mathbb{Z}/p\mathbb{Z}}(G)$ follows by induction on n in view of Lemma 1.1(ii). □

We have proved the trivial part of the next theorem, a complete proof of which is given in Passman [6].

Theorem 1.22. Let G be any group and n any natural number.

(i) (Hall-Jennings) $D_{n,Q}(G) = \sqrt{\gamma_n(G)}$.

(ii) (Jennings-Lazard-Zassenhaus) $D_{n,\mathbb{Z}/p\mathbb{Z}}(G) = M_{n,p}(G) = L_{n,p}(G)$.

It remains to evaluate the integral dimension subgroups, $D_{n,\mathbb{Z}}(G)$. It was thought for a long time that $D_{n,\mathbb{Z}}(G) = \gamma_n(G)$, the n-th term of the lower central series for G. Rips [1] was the first to give a counter example to this conjecture for $n = 4$. The result is true for $n = 2$ (Proposition 2.14), for $n = 3$ (G. Higman and Rees) and for $n = 4$ with certain exceptions (Passi [1]). Related results are given in Tahara ([1],[2]) and Sjogren [1].

2. COMMUTATIVE GROUP ALGEBRAS

In order to give an acceptable answer to the isomorphism problem in the commutative case we need the following splitting criterion of May. We say an Abelian group A splits if $A = T(A) \times B$ where $T(A)$ is the torsion subgroup of A and B is a suitable subgroup.

Proposition 2.1. An Abelian group A splits if and only if there exists a countable ascending sequence $A_1 \subseteq A_2 \subseteq \cdots \subseteq A_n \subseteq \cdots$ of subgroups of A such that

(i) $\bigcup_n A_n = A$,

(ii) $T(A_n)$ is bounded for every n i.e. there exist natural numbers m_n such that $T(A_n)^{m_n} = 1$,

(iii) $T(A/A_n) = T(A)A_n/A_n$ for every n.

Proof: See Fuchs [2], page 188.

Theorem 2.2 (May [4]). Let G be a commutative group and R an integral domain of characteristic 0. Suppose that $\mathcal{U}R \cap \{o(g): g \in G\} = \{1\}$. Then

$$RG \simeq_R RH \Rightarrow G \simeq H.$$

In order to prove this theorem we need several results. The next lemma is a special case of (II.1.4).

<u>Lemma 2.3.</u> Let G be a finite group and R an integral domain of characteristic 0 such that $\mathcal{U}R \cap \{o(g): g \in G\} = \{1\}$. Then

$$\gamma = \sum\gamma(g)g, \quad \gamma^n = 1, \quad \gamma(1) \neq 0 \Rightarrow \gamma = \gamma(1).$$

Recall that $\mathcal{U}_1(RG)$ denotes the group of 1-units of RG, namely,

$$\mathcal{U}_1(RG) = \{\alpha = \sum\alpha(g)g \in \mathcal{U}(RG): \sum\alpha(g) = 1\}.$$

<u>Corollary 2.4.</u> Let G be a finite group and R an integral domain of characteristic 0 such that $\mathcal{U}R \cap \{o(g): g \in G\} = \{1\}$. Let $\gamma \in \mathcal{U}(RG)$ be of order n. Then

(i) γ central $\Rightarrow \gamma$ is a trivial unit,

(ii) $\gamma \in \mathcal{U}_1(RG) \Rightarrow n$ is a divisor of $|G|$.

<u>Proof:</u> (i) If $\gamma = \Sigma\gamma(g)g$, $\gamma(g_0) \neq 0$ is a central unit of finite order then γg_0^{-1} is a unit of finite order and by the lemma $\gamma g_0^{-1} = \gamma(g_0)$. Thus $\gamma = \gamma(g_0)g_0$ is a trivial unit.

(ii) Let $\gamma \in \mathcal{U}_1(RG)$, $o(\gamma) = n$. Consider the idempotent

$$e = \frac{1}{n}(1 + \gamma + \cdots + \gamma^{n-1})$$

in the group algebra FG where F is the field of quotients of R. In view of Lemma 2.3, $e(1) = 1/n$. Taking the trace of the matrix E which is the image of e under the regular representation of FG, we obtain that $(G:1)1/n = m$, the rank of E. Thus $mn = (G:1)$, proving the corollary.

<u>Corollary 2.5.</u> Let G be an Abelian group and R an integral

domain of characteristic 0 with $\mathcal{U}R \cap \{o(g): g \in G\} = \{1\}$. Then $T(\mathcal{U}(RG)) = T(\mathcal{U}R) \cdot T(G)$.

Proof: We have by (II.3.5) that $\mathcal{U}(RG) = G \cdot \mathcal{U}(RT(G))$. Therefore,

$$T(\mathcal{U}(RG)) = T(G) \cdot T(\mathcal{U}(RT(G))) = T(G) \cdot T(\mathcal{U}R)$$

by Corollary 2.4.

The following is immediate from (II.3.5).

Proposition 2.6. Let G be an Abelian group and R an integral domain of characteristic 0 such that $\mathcal{U}R \cap \{o(g): g \in G\} = \{1\}$. Then

$$\mathcal{U}_1(RG) = G \cdot \mathcal{U}_1(RT(G)), \quad G \cap \mathcal{U}_1(RT(G)) = T(G).$$

Proposition 2.7. Let G be a torsion Abelian group and R an integral domain of characteristic 0 such that $\mathcal{U}R \cap \{o(g): g \in G\} = \{1\}$. Then $\mathcal{U}_1(RG)$ splits.

Proof: We shall apply May's criterion (2.1). Choose an ascending sequence

$$G_1 \subseteq G_2 \subseteq \cdots \subseteq G_n \subseteq \cdots$$

of subgroups of G such that each G_i is bounded and $G = \bigcup_i G_i$. Let

$$f_i: \mathcal{U}_1(RG) \to \mathcal{U}_1(R(G/G_i))$$

be the homomorphism induced from the natural map $RG \to R(G/G_i)$. Denote the kernel of f_i by A_i. Then, clearly,

$$A_1 \subseteq A_2 \subseteq \cdots \subseteq A_n \subseteq \cdots.$$

Also, if $x \in \mathcal{U}_1(RG)$ then there exists a j such that $x \in RG_j$ and hence $x-1 \in \Delta(G,G_j)$. It follows that $x \in A_j$ and hence

$$\mathcal{U}_1(RG) = \bigcup_i A_i .$$

Since by (2.5) $T(\mathcal{U}_1(RG)) = G$ it follows that

$$T(A_i) = \{g \in G: g-1 \in \Delta(G,G_i)\} = G_i$$

is bounded. Now, we wish to compute $T(\mathcal{U}_1(RG)/A_i)$. Observe that $u \in \mathcal{U}_1(RG)$, $u^s \in A_i \Rightarrow u^s-1 \in \Delta(G,G_i) \Rightarrow \bar{u}^s = 1$ in $R(G/G_i) \Rightarrow$ $\bar{u} = \bar{g}_0$ in $R(G/G_i)$ for some $g_0 \in G$, by (2.4) $\Rightarrow u = g_0 a$, $a \in A_i$, $g_0 \in G$. Thus $T(\mathcal{U}_1(RG)/A_i) = GA_i/A_i = T\mathcal{U}_1(RG)A_i/A_i$ and hence $\{A_i\}$ satisfies all the conditions of Proposition 2.1 thereby proving that $\mathcal{U}_1(RG)$ splits.

2.8. Proof of Theorem 2.2

Let $f: RG \to RH$ be the given normalized isomorphism. We know by (2.5) that $f: T(G) \to T(H)$ and hence $f: RT(G) \to RT(H)$. Notice that by (2.7) there exists a group Y such that

$$\mathcal{U}_1(RT(G)) = T(G) \times Y$$

which is mapped by f onto

$$\mathcal{U}_1(RT(H)) = f(T(G)) \times f(Y) = T(H) \times f(Y). \tag{2.9}$$

On the other hand by (2.6),

$$\mathcal{U}_1(RG) = G \cdot \mathcal{U}_1(RT(G)), \quad G \cap \mathcal{U}_1(RT(G)) = T(G)$$

and thus $\mathcal{U}_1(RG) = G \times Y$ which is mapped by f onto

$$\mathcal{U}_1(RH) = H \cdot \mathcal{U}_1(RT(H)), \quad H \cap \mathcal{U}_1(RT(H)) = T(H)$$
$$= H \times f(Y) \quad \text{by (2.9)}.$$

Hence we have

$$G \simeq G \times Y/Y \simeq H \times f(Y)/f(Y) \simeq H. \quad \square$$

The next corollary can also be obtained by observing that $G/G' \simeq \Delta(G)/\Delta^2(G)$ as seen in (1.15) and (1.14).

Corollary 2.10. Let G be an Abelian group. Then $\mathbb{Z}G \simeq \mathbb{Z}H \Rightarrow G \simeq H$.

The last theorem settles the isomorphism problem for Abelian groups if no group order is invertible in the coefficient ring which is an integral domain of characteristic 0. Moreover, we have

Theorem 2.11. Let RG be the group ring of an Abelian group G over an integral domain R of characteristic 0. Suppose that $RG \simeq_R RH$. Then $G/D_{2,R}(G) \simeq H/D_{2,R}(H)$.

Proof: Let Π be the set of rational primes invertible in R. Then by (1.9) and (1.14) $D_{2,R}(G)$ is the Π-torsion of G, namely, $D_{2,R}(G) = \prod_{p\in\Pi} S_p(G)$, the product of Sylow p-subgroups $S_p(G)$, $p \in \Pi$. Write $\overline{G} = G/D_{2,R}(G)$. Then it follows from (2.5) that $R\overline{G}$ has all its Π-torsion elements in R. Since $R\overline{G} \simeq RG/\Delta(G,D_{2,R}(G))$, it follows that $\Delta(G,D_{2,R}(G))$ is the smallest ideal I of RG contained in $\Delta(G)$ such that RG/I has all its Π-torsion elements in R. Hence under any isomorphism $RG \xrightarrow{f} RH$ we have

$$f(\Delta(G,D_{2,R}(G))) = \Delta(H,D_{2,R}(H)) \text{ and thus } RG/D_{2,R}(G) \simeq RH/D_{2,R}(H).$$

The result now follows by Theorem 2.2. □

The next theorem is one of the first results known on the problem.

Theorem 2.12 (Perlis and Walker [1]). If G is a finite Abelian group then

$$QG \simeq QH \Rightarrow G \simeq H.$$

Proof: We know from (II.2.6) that

$$(2.13) \qquad QG \simeq \sum^{\oplus} a_d Q(\xi_d) \simeq QH \simeq \sum^{\oplus} b_d Q(\xi_d)$$

where ξ_d is a primitive d-th root of unity, a_d (resp. b_d) denotes the number of cyclic subgroups of order d in G (resp. H). Since $Q(\xi_d) = Q$ if and only if $d = 1$ or 2 we have that $a_1 + a_2 = b_1 + b_2$ and so $a_1 = b_1$, $a_2 = b_2$. We claim

(2.14) $\quad Q(\xi_d) = Q(\xi_a) \iff d = a$ or $d = 2a$ and a odd.

Let W be the group of roots of unity in $Q(\xi_a)$ then a divides $|W|$ and since $\phi(|W|) = \phi(a)$ it follows that $|W| = a$ or $2a$, the latter if a is odd. Hence $d = a$ or $d = 2a$ and a odd. For $\alpha > 1$, $Q\left(\xi_{2^\alpha}\right) = Q(\xi_d)$ if and only if $d = 2^\alpha$ and therefore (2.13) gives that $a_{2^\alpha} = b_{2^\alpha}$. For an odd prime p, $Q\left(\xi_{p^\alpha}\right) = Q(\xi_d) \iff d = p^\alpha$ or $2p^\alpha$. Therefore,

$$b_{p^\alpha} + b_{2p^\alpha} = a_{p^\alpha} + a_{2p^\alpha} = a_{p^\alpha}(1+a_2).$$

Thus $a_{p^\alpha} = b_{p^\alpha}$ and hence the Sylow p-subgroup of G is isomorphic to the Sylow p-subgroup of H. This being true for all p, $G \simeq H$. □

For torsion free Abelian groups, of course, the isomorphism problem is trivial as seen from the next result.

<u>Theorem 2.15.</u> If G is a commutative group and K is a field, then

$$KG \simeq_K KH \Rightarrow G/T(G) \simeq H/T(H).$$

<u>Proof:</u> Notice that $KG/\Delta\big(G,T(G)\big) \simeq KG/T(G)$ being the group algebra of the torsion free commutative group $G/T(G)$ over K has all its torsion units in K. Thus $\Delta\big(G,T(G)\big)$ is the smallest ideal I contained in $\Delta(G)$ such that KG/I has all its torsion units in K. Therefore I is mapped onto its counterpart under the isomorphism and we have

$$KG/T(G) \simeq KG/\Delta\big(G,T(G)\big) \simeq KH/\Delta\big(H,T(H)\big) \simeq KH/T(H).$$

Since the 1-units of the group ring of a torsion free Abelian group over an integral domain are simply the group elements and the isomorphism $KG/T(G) \simeq KH/T(H)$ is normalized it follows that $G/T(G) \simeq H/T(H)$.

3. COMMUTATIVE MODULAR GROUP ALGEBRAS

Let G be an Abelian group and p a fixed prime. For each ordinal α, we define by induction a subgroup G^{p^α} of G. Let $G^p = \{g^p: g \in G\}$ and let G^{p^β} be defined for all ordinals $\beta < \alpha$. Then set

$$G^{p^\alpha} = \begin{cases} \left(G^{p^{\alpha-1}}\right)^p & \text{if } (\alpha-1) \text{ exists} \\ \bigcap_{\beta<\alpha} G^{p^\beta} & \text{if } \alpha \text{ is a limit ordinal.} \end{cases}$$

We denote by $G[p]$ the group $\{g \in G: g^p = 1\}$ which may be regarded as a vector space over the field F of p elements. The α-th Ulm invariant of G is the F-dimension of $G^{p^\alpha}[p]/G^{p^{\alpha+1}}[p]$. Then Ulm's Theorem (see Fuchs [2]) says that countable reduced p-groups are characterized by their Ulm invariants. Here, by a reduced group is understood a group without non-identity divisible subgroups.

Theorem 3.1 (Berman and Mollov [1], May [1]). Suppose that K is the field of p elements and G a commutative group. Then

$$KG \simeq KH \Rightarrow G \text{ and } H \text{ have the same Ulm invariants.}$$

Proof (Dubois-Sehgal [1]): Let $\theta: KG \to KH$ be the given normalized isomorphism. By noting that $KG^p = (KG)^p$ we can conclude that $\theta(KG^p) = KH^p$ and thus $\theta(KG^{p^\alpha}) = KH^{p^\alpha}$ for each ordinal α. Let us denote, for convenience, $G^{p^\alpha}[p]$ by L_α and $H^{p^\alpha}[p]$ by M_α. Then we know by (1.15) and (1.14) that

$$L_\alpha \simeq \Delta(G,L_\alpha)/\Delta(G)\Delta(G,L_\alpha)$$

the isomorphism being $x \to \overline{(x-1)}$.

Since $\Delta(G,G[p]) = \{\gamma \in KG: \gamma^p = 0\}$ we can conclude that

$$\theta: \Delta(G^{p^\alpha}[p]) \xrightarrow{\text{onto}} \Delta(H^{p^\alpha}[p])$$

and hence

$$\Delta(G,L_\alpha) = KG\ \Delta(L_\alpha) = KG\ \Delta(G^{p^\alpha}[p])$$

is mapped under θ onto $\Delta(H,M_\alpha)$. We have the commutative diagram

$$\begin{array}{ccccccc}
L_\alpha & \simeq & \Delta(G,L_\alpha)/\Delta(G)\Delta(G,L_\alpha) & \simeq & \Delta(H,M_\alpha)/\Delta(H)\Delta(H,M_\alpha) & \simeq & M_\alpha \\
\uparrow & & & & & & \uparrow \\
L_{\alpha+1} & \simeq & \Delta(G,L_{\alpha+1})/\Delta(G)\Delta(G,L_{\alpha+1}) & \simeq & \Delta(H,M_{\alpha+1})/\Delta(H)\Delta(H,M_{\alpha+1}) & \simeq & M_{\alpha+1}
\end{array}$$

and thus $L_\alpha/L_{\alpha+1} \simeq M_\alpha/M_{\alpha+1}$, proving the theorem. □

We have now the following corollary which was proved for finite groups by Deskins [1].

<u>Corollary 3.2</u>. Let G be a countable Abelian p-group and K the field of p elements. Then

$$KG \overset{\theta}{\simeq}_K KH \Rightarrow G \simeq H.$$

<u>Proof</u>: Since G is an Abelian p-group, $\Delta(G)$ and therefore $\Delta(H)$ is locally nilpotent. Thus by (I.2.21) H is a p-group. We know (see Fuchs [1]) that

$$G = D \times R, \quad H = D_1 \times R_1$$

where D and D_1 are divisible groups and R, R_1 are reduced groups. We may assume by extending θ that K is algebraically closed. Notice that $I = \{\gamma \in KG$ such that the equation $x^{p^n} = \gamma$ has a solution x in KG for every natural number $n\} = KD$. Thus

(3.3) $\qquad KD \simeq KD_1.$

Also, since the Ulm invariants for G are the same as those for R, we have by the last theorem $R \simeq R_1$. We have only to prove that $D \simeq D_1$. Since D is a divisible p-group it is a direct sum of, say, c copies of $Z(p^\infty)$ (see Fuchs [1]). In order to complete the proof it suffices to retrieve the number c from KD ring theoretically. Consider in KD the ideal $J = \{\gamma \in KD\colon \gamma^p = 0\}$ $= \Delta(D,D[p])$ where $D[p] = \{x \in D\colon x^p = 1\}$ is a direct product of c cyclic groups of order p. If c is finite $J = K(D)\Delta(D[p])$ is nilpotent of index $c(p-1) + 1$ i.e. $J^{c(p-1)+1} = 0$ but $J^{c(p-1)} \neq 0$ and if c is infinite J is not nilpotent. Since under the isomorphism of (3.3) J is mapped onto a corresponding ideal of KD_1, the number of copies of $Z(p^\infty)$ in D is the same as in D_1. Hence $D \simeq D_1$ and $G \simeq H$. □

4. FINITE NORMAL SUBGROUPS*

There are three known methods of characterizing normal subgroups of a finite group G in its integral group ring $\mathbb{Z}G$. These methods due to Berman-Passman-Glauberman, Cohn-Livingstone and Sehgal-Zmud-Kurennoi respectively generalize to different extents to cover arbitrary coefficient rings and groups. Briefly speaking, the first method consists of observing that a finite normal subgroup of G is the union of some conjugacy classes of G and that class sums are isomorphism invariants in $\mathbb{Z}G$. Whereas,

Cohn and Livingstone observed that if N is a finite normal subgroup of $\mathbb{Z}G \simeq^{\theta} \mathbb{Z}H$, one may identify $\mathbb{Z}G$ with $\mathbb{Z}H$ by replacing G by G^{θ}, and set $\phi(N) = \{h \in H: h-1 \in \Delta(G,N)\}$ and conclude that $N = \{g \in G: g-1 \in \Delta(H,\phi(N))\}$. The last method consists of observing that if N is a finite normal subgroup of G then $\hat{N} = \Sigma_{g\in N}\, g$ is a central solution of an equation $x^2 = mx$, $m \in \mathbb{Z}$ and conversely certain central solutions of this equation are, precisely, $\Sigma_{g\in N}\, g$ where N runs over finite normal subgroups of G. Some versions of the next theorem are found in Berman [2], Passman [2], Saksanov [1] and Sehgal [6].

<u>Theorem 4.1</u>. Let G be a finite group and R an integral domain of characteristic 0 such that $\mathcal{U}R \cap \{o(g): g \in G\} = \{1\}$. Let $RG \simeq^{\theta} RH$ as R-algebras. Then $\theta(C_i) = \pm K_i$ for all i, where $\{C_i\}$ and $\{K_j\}$ are class sums of G and H respectively.

<u>Proof</u>: Replacing G by G^{θ} we can assume that $RG = RH$. We have to prove that $C_i = \pm K_i$ for each i. Let us denote by C_i^* and K_i^* the conjugacy classes corresponding to C_i and K_i respectively. At first, we claim that

$$(4.2) \qquad K_i = \sum_j a_{ij} C_j, \quad a_{ij} \in \mathbb{Z}.$$

Let us denote by F the field of quotients of R and by $\{\chi_i\}$ the absolutely irreducible characters of FG. Let c_i and k_i be the number of elements in C_i^* and K_i^* respectively. Then the primitive central idempotents $\{e_i\}$ of $\mathbb{C}G$ (see Curtis and Reiner [1], page 236) are given by

$$(4.3) \qquad e_i = \frac{z_i}{(G:1)} \sum_{\nu} \overline{\chi_i(g_\nu)}\, C_\nu, \quad g_\nu \in C_\nu^*$$

where z_i is the degree of χ_i. Also,

$$(4.4) \qquad C_i = \sum_{\nu} \frac{c_i\, \chi_\nu(g_i)}{z_\nu} e_\nu, \quad g_i \in C_i^* .$$

By the same token, we have

$$(4.5) \qquad K_j = \sum_i \frac{k_j\ \chi_i(x_j)}{z_i}\, e_i, \qquad x_j \in K_j^* .$$

Substituting the value of e_i from (4.3) in (4.5) we obtain,

$$K_j = \frac{1}{(G:1)} \sum_{i,\nu} k_j\ \chi_i(x_j)\ \overline{\chi_i(g_\nu)}\ C_\nu .$$

Also,

$$K_j = \sum_\ell a_{j\ell}\ C_\ell \quad \text{for some} \quad a_{j\ell} \in R.$$

Comparing the coefficient of C_ν in the two expressions for K_j we get

$$a_{j\nu} = \frac{1}{(G:1)} \sum_i k_j\ \chi_i(x_j)\ \overline{\chi_i(g_\nu)} .$$

It follows that $a_{j\nu}$ is an algebraic number and $(G:1)a_{j\nu}$ is an algebraic integer. Let α be an imbedding of the field, generated over Q by $\{\chi_i(g),\ g \in G\}$ for all i, into the complex number field. Then χ_i^α defined by $\chi_i^\alpha(g) = (\chi_i(g))^\alpha$, $g \in G$, is also an absolutely irreducible character of G and hence $a_{j\nu}^\alpha = a_{j\nu}$. It follows that $a_{j\nu}$ is a rational number and $(G:1)a_{j\nu}$ being an algebraic integer is a rational integer. Since no divisor of (G:1) different from 1 has an inverse in R we conclude that $a_{j\nu} \in \mathbb{Z}$. This proves (4.2).

Now to class sums of every H such that RG = RH we assign a weight

$$\begin{aligned}
\omega(K_1, K_2, \cdots, K_m) &= \sum_{i,j} \chi_i(K_j)\ \overline{\chi_i(K_j)} \\
&= \sum_{i,j,\ell,s} a_{j\ell} a_{js}\ \chi_i(C_\ell)\ \overline{\chi_i(C_s)} \quad \text{(by (4.2))} \\
&= \sum_{j,\ell,s} a_{j\ell} a_{js} (G:1) c_\ell \delta_{\ell s} \quad \text{(by the orthogonality relations)} \\
&= (G:1) \sum_j \sum_\ell a_{j\ell}^2 c_\ell .
\end{aligned}$$

This weight is at least $(G:1)^2$ with equality occurring if and only if for each ℓ there exists exactly one j such that $a_{j\ell} \neq 0$ and for that j, $a_{j\ell} = \pm 1$. Hence the class sums of H have weight $(G:1)^2$ if and only if they are precisely $\{\pm C_i\}$. Reversing the roles of G and H one obtains that the class sums of a group basis have weight $(G:1)^2$ if and only if they are precisely $\{\pm K_i\}$. It follows therefore that $\{\pm C_i\} = \{\pm K_i\}$. □

<u>Corollary 4.6</u>. Let G be finite and $\mathbb{Z}G \overset{\theta}{\simeq} \mathbb{Z}H$. Then

$$N \lhd G \Rightarrow \theta(\hat{N}) = \hat{M}, \quad M \lhd H.$$

As pointed out by Ziegenbalg [1], the method of Cohn and Livingstone [2] is applicable to arbitrary normal subgroups of locally finite groups. Suppose that we have a normalized R-algebra isomorphism $\theta: RG \to RH$. We identify G^θ with G and have $RG = RH$. Let N be a normal subgroup of G. Define

$$(4.7) \qquad \phi N = \{h \in H: h-1 \in \Delta_R(G,N)\} = H \cap \left(1 + \Delta_R(G,N)\right).$$

Clearly ϕN is a normal subgroup of H. More precisely, we have

<u>Theorem 4.8</u> (Cohn-Livingstone [2] and Ziegenbalg [1]). Let G be a locally finite group. Suppose that R is an integral domain of characteristic 0 satisfying $\mathcal{U}R \cap \{o(g): g \in G\} = \{1\}$. Suppose that $RG = RH$. Then H is locally finite, $\mathcal{U}R \cap \{o(h): h \in H\} = \{1\}$ and ϕ is a 1-1, onto correspondence between $\mathcal{L}_N(G)$ and $\mathcal{L}_N(H)$, the lattices of normal subgroups of G and H respectively. Moreover, $\Delta(G,N) = \Delta(H,\phi N)$.

<u>Proof</u>: Since G is locally finite, $FG = FH$ is locally a finite dimensional algebra over F, the field of quotients of R. Thus H is locally finite. Now suppose that $h \in H$ is an element of prime order p. Then RG has an element of order p and hence by (2.4), G has an element of order p and p is not a unit of R. We have proved that $\mathcal{U}R \cap \{o(h): h \in H\} = \{1\}$. We have

defined, $\phi\colon L_N(G) \to L_N(H)$ by

$$\phi N = \{h \in H\colon h-1 \in \Delta(G,N)\}.$$

We may similarly define $\phi^*\colon L_N(H) \to L_N(G)$. In order to prove that ϕ has an inverse we have to verify that

$$N = \{g \in G\colon g-1 \in \Delta(H,\phi N)\}.$$

As $\Delta(H,\phi N) \subseteq \Delta(G,N)$ it is clear that

$$g \in G,\quad (g-1) \in \Delta(H,\phi N) \Rightarrow g \in N.$$

Conversely, we claim that

$$(4.9)\qquad n \in N \Rightarrow n-1 \in \Delta(H,\phi N).$$

Let $\{h_\nu\}$ be a transversal containing 1 of ϕN in H, $H = \bigcup_\nu h_\nu \phi N$. Then

$$(4.10)\qquad n-1 = \sum h_\nu \alpha_\nu,\quad \alpha_\nu \in R\phi N.$$

Let us apply to the last equation the natural map $\rho_N\colon RG \to RG/N$. Denoting $\rho_N(x)$ by $\bar{x}$ we get

$$0 = \sum_\nu \bar{h}_\nu \bar{\alpha}_\nu = \sum_\nu a_\nu \bar{h}_\nu$$

where $a_\nu \in R$ is the sum of the coefficients of $\alpha_\nu \in R\phi N$. Either all $a_\nu = 0$ and (4.9) follows or some a_ν, say, $a_1 \neq 0$. Then

$$-a_1 = \sum_{\nu\neq 1} a_\nu \bar{h}_\nu \bar{h}_1^{-1}$$

This is an equation in RG/N and the coefficient of identity on the left hand side is nonzero. So also the coefficient of the identity in $\bar{h}_\nu \bar{h}_1^{-1} \neq 0$ for some $\nu \neq 1$. Since H is locally finite $\bar{h}_\nu \bar{h}_1^{-1}$ is a torsion element and hence by (2.3) $\bar{h}_\nu \bar{h}_1^{-1} = 1$ and $\bar{h}_\nu = \bar{h}_1$. This means that $h_\nu h_1^{-1} \in \Delta(G,N)$ and

$h_\nu h_1^{-1} \in \phi N$. Thus $h_\nu = h_1$ which is a contradiction proving (4.9). It follows that ϕ is 1-1 and onto. It remains to prove that $\Delta(G,N) = \Delta(H,\phi N)$. The inclusion $\Delta(H,\phi N) \subseteq \Delta(G,N)$ is clear from the definition and the reverse inclusion follows from the existence of the inverse of ϕ. □

The next theorem was proved, independently, for $\mathbb{Z}G$ by Zmud-Kurennoi [1].

Theorem 4.11 (Sehgal [2],[8]). Let RG be the group ring of a group G over an integral domain R of characteristic 0 such that $\mathcal{U}R \cap \{o(g): g \in G\} = \{1\}$. Let $\gamma = \Sigma_g \gamma(g)g$ be a central element of RG satisfying

(i) $\gamma^2 = m\gamma$ for a natural number m,

(ii) $\Sigma\gamma(g) \neq 0$ and

(iii) $\gamma(1) = 1$.

Then $\gamma = \Sigma_{g \in N} g = \hat{N}$, where N is a finite normal subgroup of G.

Proof: Since $\frac{\gamma}{m}$ is a central idempotent we may suppose due to (I.4.4) that G is a finite group. Also replacing R by $\mathbb{Z}[\gamma(g): \gamma(g) \neq 0]$ we assume that R is a subring of the complex number field $\mathbb{C}$. Let us denote by $tr(\alpha)$ the trace of the $|G| \times |G|$ matrix $\Delta(\alpha)$ where $\Delta: RG \to R_{|G|}$, the ring of $|G| \times |G|$ matrices over R, is the regular representation of G with G as R-basis of RG. Since the only possible eigenvalues of $\Delta(\gamma)$ are 0 and m we have

$$(4.12) \qquad tr(\gamma) = \sum_g \gamma(g)tr(g) = (G:1)\gamma(1) = (G:1) = m\ell$$

where ℓ is the rank of $\Delta(\gamma)$.

For any g in the support of γ, γg^{-1} satisfies an equation of the type

$$x^{n+1} = m^n x.$$

Therefore, the eigenvalues of $\Delta(\gamma g^{-1})$ are 0 and $m\xi$ where ξ is an n-th root of unity. Thus diagonalizing $\Delta(\gamma)$ and $\Delta(g^{-1})$ simultaneously we see that

$$(4.13) \qquad tr(\gamma g^{-1}) = (G:1)\gamma(g) = \sum_{1}^{\ell} m\xi_i \ .$$

Let F be the cyclotomic field $Q(\xi_1, \cdots, \xi_\ell)$. Then any algebraic conjugate of $\gamma(g)$ in F is of the form

$$(4.14) \qquad \gamma(g)' = \frac{1}{(G:1)} \sum_{1}^{\ell} m\xi_i^r$$

for a suitable r. But this is the first coefficient of γg^{-r} as seen by taking the trace of $\Delta(\gamma g^{-r})$. Hence $\gamma(g)'$ and $a = \operatorname{Norm}_{F/Q}(\gamma(g)) \in R$. Taking $\operatorname{Norm}_{F/Q}$ of both sides of (4.13) we have

$$(4.15) \qquad (G:1)^s a = \text{a sum of } (m\ell)^s \text{ roots of unity}$$

where $s = (F:Q)$ and $a \in Q \cap R$. Now $(G:1)^s a$ is an algebraic integer and also a rational number. Therefore, $(G:1)^s a$ is a rational integer and since no divisor of $(G:1)$ other than one is invertible in R it follows that a is a rational integer. Remembering that $m\ell = (G:1)$ as seen in (4.12) and taking absolute value of both sides of (4.15) we can conclude that if $\gamma(g) \neq 0$ then all the roots of unity on the right hand side of (4.15) are equal. It is easy to see then that all ξ_i's in (4.13) are equal say to ξ_g. We conclude from (4.13) and (4.12) that

$$\gamma(g) = \xi_g \quad \text{where} \quad \xi_g^n = 1.$$

Write $\gamma^* = \Sigma_g \overline{\gamma(g)}\, g^{-1}$ where $\overline{}$ is the complex conjugate. Then

$$(\gamma\gamma^*)^2 = m^2(\gamma\gamma^*).$$

Taking trace of the regular representation matrix $\Delta(\gamma\gamma^*)$ we have

$$tr(\gamma\gamma^*) = (G:1)\gamma\gamma^*(1) = (G:1)\sum \xi_g \overline{\xi}_g \le m^2 \ell = m(G:1)$$

and therefore,

$$\sum \xi_g \overline{\xi}_g \leq m.$$

Hence the number of nonzero terms in $\gamma = \Sigma\gamma(g)g$ is at most m. But $\gamma^2 = m\gamma$ implies that $\Sigma\gamma(g) = m$, therefore,

$$\sum \xi_g = m \tag{4.16}$$

where the number of nonzero terms on the left hand side of (4.16) is at most m. This means that each ξ_g is equal to 1. Hence $\gamma = \Sigma_{g\in N}\, g$. Now (i) implies that N is a group and it is normal as γ is central. □

<u>Theorem 4.17</u>. Let R be an integral domain of characteristic 0 in which no rational prime number has an inverse. Suppose that $RG \cong_R RH$. Then there exists a 1-1, onto correspondence $N \leftrightarrow M$ between the lattice $L_{FN}(G)$ of finite normal subgroups of G and that of H.

<u>Proof</u>: Let $\theta\colon RG \to RH$ be the normalized R-isomorphism. Let $\gamma = \Sigma_{g\in N}\, g = \hat{N}$ for a finite normal subgroup N of G. Then γ is central and satisfies $\gamma^2 = m\gamma$, $m = |N|$, $\gamma(1) = 1$. Also, $\theta(\gamma) = \mu = \Sigma_{h\in H}\mu(h)h$ satisfies $x^2 = mx$, is central and has augmentation $\Sigma\mu(h) = m$. Writing $\theta(g) = \Sigma_{h\in H}\, b_h h$ it follows from (II.1.2) that $b_1 = 0$ for $g \neq 1$. Thus μ has the property that $\mu(1) = 1$. We conclude by (4.11) that $\theta(\gamma) = \mu = \Sigma_{h\in M}\, h$ for a finite normal subgroup M of H. We have obtained a 1-1, onto correspondence $N \leftrightarrow M$. □

Let us denote by $A(S)$ the left annihilator of a subset S of a ring R, namely,

$$A(S) = \{r \in R\colon rs = 0 \text{ for all } s \in S\}.$$

The following is a useful fact.

Proposition 4.18. Let N be a normal subgroup of G.

(1) If N is infinite, $A(\Delta_R(G,N)) = 0$,

(2) If N is finite, $A(\Delta_R(G,N)) = (RG)\hat{N}$ and $A(\hat{N}) = \Delta_R(G,N)$.

Proof: Suppose that $\gamma = \Sigma_{g \in G} \gamma(g)g$ satisfies $\gamma(1-n) = 0$ for all $n \in N$. Then $\gamma = \gamma n$ and consequently $\gamma(g) = \gamma(gn)$ for all $n \in N$. If N is infinite this implies that $\gamma(g) = 0$ for all g because the support of γ is finite. If N is finite then $\gamma = \mu\hat{N}$ for a suitable $\mu \in RG$.

Now suppose that $\gamma\hat{N} = 0$. Choose a transversal T for N in G. Then

$$\gamma = \sum_{t \in T} \beta_t\, t, \quad \beta_t \in RN \quad \text{and} \quad 0 = \gamma\hat{N} = \sum_{t \in T} \beta_t\, \hat{N}t.$$

Thus $\beta_t\hat{N} = 0$ for all t. Consequently $\beta_t \in \Delta_R(N)$ and $\gamma \in \Delta_R(G,N)$. □

Remark 4.19. Let N be a finite normal subgroup of G such that $RG = RH$ and $N \leftrightarrow M$ as in (4.17). Then $\hat{N} = \hat{M}$ and $\Delta(G,N) = \Delta(H,M)$ by the last proposition. Thus

$$\phi N = \{h \in H\colon (h-1) \in \Delta(G,N)\} = M$$

and ϕ as defined in (4.7) coincides with $\leftrightarrow$.

To establish the properties of this correspondence we need a couple of useful results which are interesting in their own right. Let N be a normal subgroup of G. Then the factor group N/N', where N' is the derived group of N, can be regarded as a $\mathbb{Z}G$-module by setting

$$(\bar{n})^g = \overline{g^{-1}ng}, \quad \bar{n} \in N/N', \quad g \in G$$

and linear extension. Also $\Delta(G,N)$ is a right $\mathbb{Z}G$-module by multiplication. We have the following proposition which is an extension of (1.15). This result goes way back but has recently been rediscovered by several authors.

Proposition 4.20. Let N be a normal subgroup of G. Then

$$N/N' \simeq \Delta_{\mathbb{Z}}(G,N)/\Delta_{\mathbb{Z}}(G)\Delta_{\mathbb{Z}}(G,N)$$

as $\mathbb{Z}G$-modules.

Proof: It has been proved in (1.15) that

$$\theta\colon N/N' \to \Delta_{\mathbb{Z}}(G,N)/\Delta_{\mathbb{Z}}(G)\Delta_{\mathbb{Z}}(G,N)$$

given by $\theta(\bar{n}) = \overline{n-1}$ is an isomorphism of additive groups. Since

$$\theta(\bar{n}^g) = \theta(\overline{g^{-1}ng}) = \overline{g^{-1}ng - 1} = \overline{g^{-1}(n-1)g} = \overline{(n-1)}g$$

it follows that θ is an isomorphism of $\mathbb{Z}G$-modules. □

The next result is due to Passman [2], Sandling [5] and Whitcomb [1].

Proposition 4.21. Let RG be the group ring of a group G over an integral domain R of characteristic 0 such that $\mathcal{U}R \cap \{o(g)\colon g \in G\} = \{1\}$. Suppose that M and N are two normal subgroups of G. Let $[\Delta_R(G,N),\Delta_R(G,M)]$ be the R-module generated by all $[\alpha,\beta] = \alpha\beta - \beta\alpha$, $\alpha \in \Delta_R(G,N)$, $\beta \in \Delta_R(G,M)$. Then

$$(N,M) = G \cap \left(1 + [\Delta_R(G,N),\Delta_R(G,M)]RG\right) .$$

Proof: By considering $G/(N,M)$ we may suppose that $(N,M) = 1$. Let

$$g-1 \in [\Delta(G,N),\Delta(G,M)]RG, \quad g \in G.$$

Then since

$$[\Delta(G,N),\Delta(G,M)] \subseteq \Delta(N)\Delta(M)RG \subseteq \Delta(G,NM) \tag{4.22}$$

it follows that $g \in NM$. Pick a transversal, $T = \{t_\nu\}$ containing 1, of NM in G. Write any element x of G as $x = bt_\nu$, $b \in NM$ and define a map $\rho\colon G \to NM$ by $\rho(x) = b$.

Extend this map R-linearly to $\rho: RG \to R(NM)$. Applying ρ to the equation

$$g-1 = \sum r_{ij}(1-n_i)(1-m_j)t_{ij}, \quad r_{ij} \in R, \quad n_i \in N, \quad m_j \in M, \quad t_{ij} \in T,$$

which is a consequence of the assumption on g, we conclude that

(4.23) $\quad g-1 \in \Delta(N)\Delta(M)$.

Thus $g \in N \cap M = A$ which, of course, is central in NM. Let S and L (both containing 1) be transversals of A in N and M respectively. Then SL is a transversal of A in NM. Now apply the map $g \to a$ extended R-linearly, if g can be written as $g = as\ell$, $a \in A$, $s \in S$, $\ell \in L$, to the equation

$$g-1 = \sum_{i,j} \alpha_{ij}(a_i s_i - 1)(a_j' \ell_j - 1), \quad a_i, a_j' \in A, \quad s_i \in S, \quad \ell_j \in L, \quad \alpha_{ij} \in R$$

which is implied by (4.23). We obtain

$$g-1 = \sum_{i,j} \alpha_{ij}(a_i - 1)(a_j' - 1) \in \Delta^2(A).$$

Hence $g \in D_{2,R}(A)$ which is $\{1\}$ by Theorem 1.9 and Proposition 1.14 proving the result. □

Proposition 4.24 (Bergman and Dicks [1]). Let RG be the group ring of a group G over an integral domain of characteristic 0 such that $\mathcal{U}R \cap \{o(g): g \in G\} = \{1\}$. Suppose that M and N are two normal subgroups of G. Then

$$(N \cap M)' = G \cap 1 + \Delta(G,N)\Delta(G,M).$$

Proof: By considering $G/(N \cap M)'$ if necessary we may suppose that $(N \cap M)' = 1$. Let

(4.25) $\quad g-1 \in \Delta(G,N)\Delta(G,M) = \Delta(N)\Delta(M)RG, \quad g \in G.$

Since the right hand side is contained in $\Delta(G,NM)$ it follows that $g \in NM$. By picking a transversal for NM in G and applying a suitable map to the equation resulting from (4.25) as in the last proposition we conclude that

$$g-1 \in \Delta(NM,N)\Delta(NM,M).$$

In other words, we may take $G = NM$ with $(N \cap M)' = \{1\}$. Letting $A = N \cap M$, we pick S and T transversals containing 1 of A in N and M respectively. Then each element of G is expressible uniquely as a product sat, $s \in S$, $a \in A$, $t \in T$. Define $\rho\colon RG \to RA$ by $\rho(sat) = a$ and linear extension.

It is easy to check that $\Delta(NM,N)\Delta(NM,M) = \Delta(N)\Delta(M)$ which implies that $g \in A$ and

$$g-1 = \sum_{i,j} \alpha_{ij}(1-n_i)(1-m_j), \quad n_i \in N, \quad m_j \in M, \quad \alpha_{ij} \in R.$$

Let $n_i = s_i a_i$, $m_j = a_j' t_j$, $s_i \in S$, $t_j \in T$, $a_i, a_j' \in A$. Then

$$\begin{aligned} g-1 = \rho(g-1) &= \sum \alpha_{ij}\, \rho(1-n_i-m_j+n_i m_j) \\ &= \sum \alpha_{ij}(1-a_i-a_j'+a_i a_j') \\ &= \sum \alpha_{ij}(1-a_i)(1-a_j') \in \Delta(A)^2. \end{aligned}$$

Hence $g \in D_{2,R}(A)$ which is $\{1\}$ by Theorem 1.9 and Proposition 1.14. □

<u>Theorem 4.26</u>. Let R be an integral domain of characteristic 0 such that no rational prime is invertible in R. Suppose that $RG = RH$. Let $\phi\colon L_{FN}(G) \to L_{FN}(H)$ be the 1-1 correspondence between the lattices of finite normal subgroups of G and H given by

$$G \trianglerighteq N \leftrightarrow \phi N = \{h \in H\colon h-1 \in \Delta(G,N)\} \subseteq H.$$

Then

(i) $\Delta(G,N) = \Delta(H,\phi N)$, $RG/N \simeq RH/\phi N$,

(ii) $N_1 \subseteq N_2 \Rightarrow \phi N_1 \subseteq \phi N_2$,

(iii) If I is an ideal of RG and N is the normal subgroup $N = G \cap (1+I)$ then $\phi N = H \cap (1+I)$,

(iv) For $M,N \in L_{FN}(G)$, $\phi(M,N) = (\phi M,\phi N)$ and $\phi(M \cap N)' = (\phi M \cap \phi N)'$.

Proof: (i) is Remark (4.19) and (ii) follows from the fact that $N_1 \subseteq N_2 \Rightarrow \Delta(G,N_1) \subseteq \Delta(G,N_2)$.

(iii) $\phi N = \{h \in H\colon h-1 \in \Delta(G,N)\} \subseteq \{h \in H\colon h-1 \in I\} = H \cap (1+I)$.

The equality, $\phi N = H \cap (1+I)$, now follows from the invertibility of ϕ.

(iv) is a consequence of (4.21) and (4.24) in view of (iii) and (i), for example,

$$\begin{aligned}\phi(N \cap M)' &= \phi\Big(G \cap \big(1 + \Delta(G,N)\Delta(G,M)\big)\Big)\\ &= H \cap \big(1 + \Delta(G,N)\Delta(G,M)\big)\\ &= H \cap \big(1 + \Delta(H,\phi N)\Delta(H,\phi M)\big)\\ &= (\phi N \cap \phi M)'. \quad \square\end{aligned}$$

The same argument gives us the

Theorem 4.27. Suppose that G is a locally finite group and R is an integral domain of characteristic 0 such that $UR \cap \{o(g)\colon g \in G\} = \{1\}$ with $RG \simeq RH$. Then $\phi\colon L_N(G) \to L_N(H)$ the 1-1 correspondence between the lattices of normal subgroups of G and H as defined earlier, namely,

$$G \supseteq N \leftrightarrow \phi N = \{h \in H\colon h-1 \in \Delta(G,N)\} \subseteq H$$

satisfies (i)-(iv) of the last theorem.

It follows from the above theorem that for locally finite groups G the nilpotence class and solvability length of G are invariants of $\mathbb{Z}G$. Let us first recall that $\{\delta_i(G)\}$, $\{\gamma_i(G)\}$, $\{\zeta_i(G)\}$ denote the derived, lower central and upper central series of G. These are defined inductively:

$$\delta_1(G) = (G,G),\ \gamma_1(G) = G,\ \zeta_1(G) = \text{the center of } G = \zeta(G)$$

and for $i \geq 1$

$$\delta_{i+1}(G) = \big(\delta_i(G),\delta_i(G)\big),\gamma_{i+1}(G) = \big(\gamma_i(G),G\big),\zeta_{i+1}(G)/\zeta_i(G) = \zeta\big(G/\zeta_i(G)\big).$$

We have the following theorem.

Theorem 4.28. Suppose that $\mathbb{Z}G = \mathbb{Z}H$ for a locally finite group G. Let ϕ be the normal subgroup correspondence $L_N(G) \to L_N(H)$. Then for all $i \geq 1$,

$$\phi\big(\delta_i(G)\big) = \delta_i(H),\ \phi\big(\gamma_i(G)\big) = \gamma_i(H),\ \phi\big(\zeta_i(G)\big) = \zeta_i(H) \quad \text{and}$$

$$\delta_i(G)/\delta_{i+1}(G) \simeq \delta_i(H)/\delta_{i+1}(H),\ \gamma_i(G)/\gamma_{i+1}(G) \simeq \gamma_i(H)/\gamma_{i+1}(H),$$

$$\zeta_{i+1}(G)/\zeta_i(G) \simeq \zeta_{i+1}(H)/\zeta_i(H).$$

We need the following

Lemma 4.29. For a normal subgroup N of G we have

$$N/(G,N) \simeq \Delta_{\mathbb{Z}}(G,N)/\big(\Delta_{\mathbb{Z}}(G)\Delta_{\mathbb{Z}}(G,N) + \Delta_{\mathbb{Z}}(G,N)\Delta_{\mathbb{Z}}(G)\big).$$

Proof: Consider the map

$$\theta\colon N \to \Delta_{\mathbb{Z}}(G,N)/\big(\Delta_{\mathbb{Z}}(G)\Delta_{\mathbb{Z}}(G,N) + \Delta_{\mathbb{Z}}(G,N)\Delta_{\mathbb{Z}}(G)\big)$$

given by $\theta(n) = \overline{(n-1)}$ for $n \in N$. Clearly θ is a homomorphism as $n_1n_2 - 1 = (n_1-1) + (n_2-1) + (n_1-1)(n_2-1)$. And θ is onto as $g(n-1) = n - 1 + (g-1)(n-1)$. It remains to prove:

$$n \in N,\ n-1 \in \Delta_{\mathbb{Z}}(G)\Delta_{\mathbb{Z}}(G,N) + \Delta_{\mathbb{Z}}(G,N)\Delta_{\mathbb{Z}}(G) \Rightarrow n \in (G,N).$$

By considering $G/(G,N)$, we may suppose that $(G,N) = 1$ which implies that $\Delta(G)\Delta(G,N) = \Delta(G,N)\Delta(G)$ and we have to prove:

$$n \in N, \quad n-1 \in \Delta_{\mathbb{Z}}(G)\Delta_{\mathbb{Z}}(G,N) \Rightarrow n = 1.$$

But this statement follows from (1.19) together with the fact that $N' \subseteq (G,N) = \{1\}$. □

4.30. Proof of Theorem 4.28

It follows from the definition of ϕ that $\phi G = H$ and from (iv) of the last theorem that $\phi(\delta_i(G)) = \delta_i(H)$, $\phi(\gamma_i(G)) = \gamma_i(H)$ for all i. From the isomorphisms

$$N/N' \simeq \Delta(G,N)/\Delta(G)\Delta(G,N)$$

and

$$N/(G,N) \simeq \Delta(G,N)/[\Delta(G)\Delta(G,N) + \Delta(G,N)\Delta(G)]$$

established in (1.15) and (4.29) we can conclude that

$$\delta_i(G)/\delta_{i+1}(G) \simeq \delta_i(H)/\delta_{i+1}H \quad \text{and} \quad \gamma_i(G)/\gamma_{i+1}(G) \simeq \gamma_i(H)/\gamma_{i+1}(H).$$

We know (2.4) that if we denote by TA the torsion elements of a group A,

$$\zeta_1(G) = \zeta(G) = T\zeta(\mathcal{U}_1(\mathbb{Z}G)) = T\zeta(\mathcal{U}_1(\mathbb{Z}H)) = \zeta_1(H)$$

and thus

$$\phi(\zeta_1(G)) = \{h \in H\colon h-1 \in \Delta(G,\zeta_1(G)) = \Delta(H,\zeta_1(H))\} = \zeta_1(H).$$

We have therefore

$$\mathbb{Z}(G/\zeta_1(G)) \simeq \mathbb{Z}G/\Delta(G,\zeta_1(G)) = \mathbb{Z}H/\Delta(H,\zeta_1(H)) \simeq \mathbb{Z}H/\zeta_1(H)$$

and applying (2.4) to $\mathbb{Z}(G/\zeta_1(G))$ we get $\zeta_2(G)/\zeta_1(G) \simeq \zeta_2(H)/\zeta_1(H)$. Let us assume, by induction, $\phi(\zeta_t(G)) = \zeta_t(H)$, $\zeta_t(G)/\zeta_{t-1}(G) \simeq \zeta_t(H)/\zeta_{t-1}(H)$ for all $t \le i$. Then $\phi(\zeta_{i+1}(G)) = \{h \in H\colon h-1 \in \Delta(G,\zeta_{i+1}(G))\} = M$, say. For $h \in H$, $m \in M$ we have by (4.27)

$$(m,h) = m^{-1}h^{-1}mh \in (M,H) = \phi(\zeta_{i+1}(G),G).$$

But $(\zeta_{i+1}(G),G) \subseteq \zeta_i(G)$ and accordingly $(m,h) \in \phi(\zeta_i(G)) = \zeta_i(H)$. We have proved that $(m,h) \in \zeta_i(H)$ for all $h \in H$ which is to say that $m \in \zeta_{i+1}(H)$. Therefore, $\phi(\zeta_{i+1}(G)) \subseteq \zeta_{i+1}(H)$ and equality follows from the invertibility of ϕ. Now the observation, $\zeta_{i+1}(G)/\zeta_i(G)$

$$= \zeta(G/\zeta_i(G)) = T\zeta\left(\mathcal{U}_1(\mathbb{Z}G/\zeta_i(G))\right) \simeq T\zeta\left(\mathcal{U}_1(\mathbb{Z}G/\Delta(G,\zeta_i(G)))\right)$$

$$= T\zeta\left(\mathcal{U}_1(\mathbb{Z}H/\Delta(H,\zeta_i(H)))\right) \simeq T\zeta\left(\mathcal{U}_1(\mathbb{Z}H/\zeta_i(H))\right) = \zeta_{i+1}(H)/\zeta_i(H)$$

completes the proof of the theorem. □

5. FINITE METABELIAN GROUPS

Whitcomb [1] proved that finite metabelian groups are characterized by their integral group rings. Jackson [1] gave an incomplete proof whereas Obayashi [1] gave a homological proof of the same result. We need (1.19), (II.1.7) and (2.4) which we state below.

Lemma 5.1. $g \in G$, $g-1 \in \Delta_R(G)\Delta_R(G,N)$, N normal subgroup of $G \Rightarrow g \in D_{2,R}(N)$.

Lemma 5.2. (i) Any central unit γ of finite order in the integral group ring $\mathbb{Z}G$ is trivial.

(ii) Any central unit γ of finite order in the group ring RG, of a finite group over an integral domain of characteristic 0 satisfying $\mathcal{U}R \cap \{o(g): g \in G\} = \{1\}$, is trivial.

Theorem 5.3 (Whitcomb [1]). Let G be a torsion group. Then

$$\mathbb{Z}G \simeq \mathbb{Z}H \Rightarrow G/G'' \simeq H/H''.$$

Proof: Let $\theta: \mathbb{Z}G \to \mathbb{Z}H$ be a normalized isomorphism. Since $\Delta(G,G')$ is the smallest ideal I such that $\mathbb{Z}G/I$ is commutative, we have that $\theta\Delta(G,G') = \Delta(H,H')$.

Let $g \in G$ and $\theta(g) = \gamma$. Then $\overline{\gamma} \in \mathbb{Z}H/H'$ is a unit of finite order in the group ring of a suitable finite Abelian group and thus by (5.2) $\overline{\gamma} = \overline{h}_0$, $h_0 \in H$. In other words,

$$\begin{aligned}\gamma &= h_0 + \delta, \quad \delta \in \Delta(H,H'), \\ &= h_0 + \sum_x \alpha_x(x-1), \quad x \in H', \quad \alpha_x \in \mathbb{Z}H, \\ &\equiv h_0 + \sum c_x(x-1) \bmod \Delta(H)\Delta(H,H') \quad \text{where } c_x = c(\alpha_x) \in \mathbb{Z}, \\ &\equiv 1 + (h_0-1) + \sum c_x(x-1) \bmod \Delta(H)\Delta(H,H').\end{aligned}$$

By using the equation

$$ab-1 = (a-1) + (b-1) + (a-1)(b-1)$$

and its consequence,

$$a^{-1} - 1 = -(a-1) + (a-1)(a^{-1}-1)$$

we can conclude that

$$\gamma-1 \equiv h_0(\Pi x^{c_x}) - 1 \bmod \Delta(H)\Delta(H,H').$$

Thus if

$$\theta(g) \equiv h_g \bmod \Delta(H)\Delta(H,H'), \quad h_g \in H$$

we define a map

$$G/G'' \ni \overline{g} \xrightarrow{\beta} \overline{h}_g \in H/H''.$$

β is well defined as $h_1 \equiv h_2 \bmod \Delta(H)\Delta(H,H')$ implies that $h_1h_2^{-1} \equiv 1 \bmod \Delta(H)\Delta(H,H')$ and that $h_1h_2^{-1} \in H''$ by (5.1). Clearly β is a homomorphism. Suppose that $\beta(\overline{g}) = \overline{h}$, $h \in H''$. Then $\theta(g) \equiv h \equiv 1 \bmod \Delta(H)\Delta(H,H')$ and therefore $g \equiv 1 \bmod \Delta(G)\Delta(G,G')$ and $g \in G''$ by (5.1). Thus β is one to one. It remains to prove that β is onto. Given $h \in H$, there exists a $\mu \in \mathbb{Z}G$ such that $\theta(\mu) = h$. Then $\overline{\mu} \in \mathbb{Z}G/G'$ is $\overline{g}_0$

for some $g_0 \in G$ and as before,

$$\mu \equiv g \bmod \Delta(G)\Delta(G,G'), \quad \text{for some} \quad g \in G.$$

Hence $\theta(g) \equiv \theta(\mu) \equiv h \bmod \Delta(H)\Delta(H,H')$ and $\beta(\bar{g}) = \bar{h}$ proving that β is onto and $G/G'' \simeq H/H''$. □

<u>Corollary 5.4</u>. G torsion metabelian, $\mathbb{Z}G \simeq \mathbb{Z}H \Rightarrow G \simeq H$.

<u>Proof</u>: Theorems 4.28 and 5.3.

<u>Corollary 5.5</u> (Higman and Passman). G torsion, nilpotent of clas two, $\mathbb{Z}G \simeq \mathbb{Z}H \Rightarrow G \simeq H$.

The last corollary also follows from

<u>Theorem 5.6</u>. G, H torsion, $\mathbb{Z}G \overset{\theta}{\simeq} \mathbb{Z}H \Rightarrow \zeta_2(G) \simeq \zeta_2(H)$.

<u>Proof</u>: We notice that by (5.2) $\theta\zeta_1(G) = \zeta_1(H)$. Let $g \in \zeta_2(G)$. Then $g\zeta_1(G)$ is a central unit of finite order in $\mathbb{Z}(G/\zeta_1(G))$ an hence $\theta(g)\zeta_1(H)$ is also a central unit of finite order in $\mathbb{Z}(H/\zeta_1(H))$. Hence by (5.2) $\theta(g)\zeta_1(H) = h\zeta_1(H)$ for some $h \in \zeta_2(H)$. We have

$$\theta(g) = h + \delta, \quad \delta \in \Delta(H,\zeta_1(H))$$

and consequently

$$\theta(g) \equiv h_g \bmod \Delta(H)\Delta(H,\zeta_1(H)), \quad h_g \in \zeta_2(H).$$

As in (5.3) one can prove now that

$$\beta\colon g \to h_g$$

is an isomorphism of $\zeta_2(G)$ onto $\zeta_2(H)$. □

Next we wish to prove the same results for the p-adic group rings of finite p-groups. This is useful because, as we shall see, isomorphisms of certain modular group rings can be lifted to isomorphisms of p-adic group rings. Let us denote by $\mathbb{Z}_p$ (Q_p) the ring (field) of p-adic integers (numbers). We first make an observation. Let $g \in G$ satisfy $g^n = 1$. Then

$$0 = g^n - 1 = \big(1 + (g-1)\big)^n - 1 = n(g-1) + \binom{n}{2}(g-1)^2 + \cdots + (g-1)^n$$

and $n(g-1) \in (g-1)^2 RG$. As a result we have proved

(5.7) Let N be a finite normal subgroup of G. Then $(N:1)\Delta(N) \subseteq \Delta(G,N)^2$.

<u>Theorem 5.8</u> (Sehgal [5]). Let G be a finite metabelian p-group. Then

$$\mathbb{Z}_p G \simeq^{\theta} \mathbb{Z}_p H \Rightarrow G \simeq H.$$

<u>Proof</u>: Since $\mathbb{Z}_p G \simeq \mathbb{Z}_p H \Rightarrow \mathbb{Z}/p\mathbb{Z}G \simeq \mathbb{Z}_p G/p\mathbb{Z}_p G \simeq \mathbb{Z}_p H/p\mathbb{Z}_p H \simeq \mathbb{Z}/p\mathbb{Z}H$ we conclude that $(G:1) = (H:1)$. Also H is a metabelian p-group in view of (4.27). For $g \in G$, let $\theta(g) = \gamma$. Then by (5.2) in $\mathbb{Z}_p(H/H')$, $\overline{\gamma} = \overline{h}_0$ for some $h_0 \in H$. Thus

$$\begin{aligned} \gamma &= h_0 + \delta, \quad \delta \in \Delta_{\mathbb{Z}_p}(H,H'), \\ &= h_0 + \delta_1 + p^m\delta_2, \quad \delta_1 \in \Delta_{\mathbb{Z}}(H,H'), \quad \delta_2 \in \Delta_{\mathbb{Z}_p}(H,H'), \quad p^m = (H:1) \\ &\equiv h_0 + \delta_1 \bmod \Delta^2_{\mathbb{Z}_p}(H,H') \quad \text{by (5.7)}. \end{aligned}$$

As in (5.3) we can deduce that

$$\theta(g) \equiv h_g \bmod \Delta_{\mathbb{Z}_p}(H)\Delta_{\mathbb{Z}_p}(H,H')$$

for some $h_g \in H$. The map $G \to H$ given by $g \to h_g$ is well defined and 1-1 can be seen by (5.1) using the fact (1.9) that $D_{2,\mathbb{Z}_p}(P) = P'$ for any p-group P. The map is onto as in (5.3) and hence $G \simeq H$. □

The same arguments also give

Theorem 5.9. Let G be a finite p-group. Then

$$\mathbb{Z}_pG \simeq \mathbb{Z}_pH \Rightarrow G/G'' \simeq H/H'', \quad \zeta_2(G) \simeq \zeta_2(H).$$

The next theorem is a special case of a result of D.G. Higman [1]. See also Sehgal [5]. In the following it is convenient to denote by (a) the ideal aR of a ring R.

Theorem 5.10. Let G be a group of order $p^m s$ where $(p,s) = 1$. Suppose that $\theta: \mathbb{Z}/(p^{2m+1})G \to \mathbb{Z}/(p^{2m+1})H$ is an isomorphism. Then $\mathbb{Z}_pG \simeq \mathbb{Z}_pH$.

Proof: Since the hypothesis implies $\mathbb{Z}/(p)G \simeq \mathbb{Z}/(p)H$ it follows that $|H| = |G| = p^m s$. We wish to construct, by induction, a sequence of maps $\alpha_i: G \to \mathbb{Z}_pH$ such that

$$\text{(i)}\ \alpha_i(xy) \equiv \alpha_i(x)\alpha_i(y) \bmod p^{2m+i}$$

and

$$\text{(ii)}\ \alpha_{i+1}(x) \equiv \alpha_i(x) \bmod p^{m+i}$$

for all i and all $x,y \in G$. Since $\mathbb{Z}/(p^{2m+1}) \simeq \mathbb{Z}_p/(p^{2m+1})$, given $x \in G$, we pick a representative for $\theta(x)$ in $\mathbb{Z}_pH$ and set it equal to $\alpha_1(x)$. Suppose that α_i is defined for $i \geq 1$ and we shall define α_{i+1}. Let G act on $\mathbb{Z}_p/(p^{2m+i})H$ by

$$x \cdot a = \alpha_i(x)a, \quad a \cdot x = a\alpha_i(x)$$

for $x \in G$ and $a \in \mathbb{Z}_p/(p^{2m+i})H$. Write

$$f(x,y) = \alpha_i(xy) - \alpha_i(x)\alpha_i(y) = p^{2m+i}s\ \beta(x,y).$$

Then,

$$(\partial\beta)(x,y,z) = x\beta(y,z) - \beta(xy,z) + \beta(x,yz) - \beta(x,y)z = 0.$$

Thus β, modulo p^{2m+i}, is a 2-cocycle i.e. $\beta \in Z^2(G,\mathbb{Z}_p/(p^{2m+i})H)$. Since the second cohomology group $H^2(G,Z_p/(p^{2m+i})H)$ is of exponent dividing $|G|$ (see M. Hall [1]) $p^m s\beta$ equals $\partial\lambda$ for some 1-cochain $\lambda \in C^1(G,\mathbb{Z}_p/(p^{2m+i})H)$. We define $\tilde{\lambda}\colon G \to \mathbb{Z}_pH$ by setting for every $g \in G$, $\tilde{\lambda}(g)$ equal to a representative of $\lambda(g)$ in $\mathbb{Z}_pH$. Let

$$\alpha_{i+1}(x) = \alpha_i(x) + p^{m+i}\tilde{\lambda}(x).$$

Then, $\alpha_{i+1}(xy)$

$$\begin{aligned}
&= \alpha_i(xy) + p^{m+i}\tilde{\lambda}(xy) \\
&= \alpha_i(x)\alpha_i(y) + p^{2m+i}s\ \beta(x,y) + p^{m+i}\tilde{\lambda}(xy) \\
&\equiv \alpha_i(x)\alpha_i(y) + p^{m+i}\big(\partial\tilde{\lambda}(x,y) + \tilde{\lambda}(xy)\big) \pmod{p^{3m+2i}} \\
&\equiv \alpha_i(x)\alpha_i(y) + p^{m+i}\big(x\cdot\tilde{\lambda}(y) - \tilde{\lambda}(xy) + \tilde{\lambda}(x)\cdot y + \tilde{\lambda}(xy)\big) \pmod{p^{3m+2i}} \\
&\equiv \alpha_i(x)\alpha_i(y) + p^{m+i}\big(\alpha_i(x)\tilde{\lambda}(y) + \tilde{\lambda}(x)\alpha_i(y)\big) \pmod{p^{3m+2i}}
\end{aligned}$$

and $\alpha_{i+1}(x)\alpha_{i+1}(y)$

$$\begin{aligned}
&= \big(\alpha_i(x) + p^{m+i}\tilde{\lambda}(x)\big)\big(\alpha_i(y) + p^{m+i}\tilde{\lambda}(y)\big) \\
&\equiv \alpha_i(x)\alpha_i(y) + p^{m+i}\big(\alpha_i(x)\tilde{\lambda}(y) + \tilde{\lambda}(x)\alpha_i(y)\big) \pmod{p^{2m+2i}}.
\end{aligned}$$

Thus

$$\alpha_{i+1}(xy) \equiv \alpha_{i+1}(x)\alpha_{i+1}(y) \pmod{p^{2m+2i}}.$$

Now let $\alpha(x) = \lim_{i\to\infty} \alpha_i(x)$ (here, limit is taken at each component). Then $\alpha(xy) = \alpha(x)\alpha(y)$ for all $x,y \in G$. Extend α linearly to $\alpha\colon \mathbb{Z}_pG \to \mathbb{Z}_pH$. We claim that α is onto. Given $a \in \mathbb{Z}_pH$, there exists b_0 in $\mathbb{Z}_pG$ such that $\alpha(b_0) = a + p^{2m+1}a_1$. Again, there exists a $b_1 \in \mathbb{Z}_pG$ such that $\alpha(b_1) = a_1 + p^{2m+1}a_2$ and thus

$$\alpha(b_0 - p^{2m+1}b_1) = a - p^{2(2m+1)}a_2\ .$$

Continuing in this manner we get a sequence $\{b_i\}$ of elements in $\mathbb{Z}_pG$ such that $b = b_0 - \Sigma p^{(2m+1)i} b_i$ is mapped under α onto a. Since α can be extended to an onto map $Q_pG \to Q_pH$, α is also 1-1. Hence we have proved that $\mathbb{Z}_pG \simeq \mathbb{Z}_pH$. □

Remark 5.11. It would be nice to know if it is true that

$$\mathbb{Z}/p\mathbb{Z}G \simeq \mathbb{Z}/p\mathbb{Z}H, \quad G \text{ finite p-group} \Rightarrow \mathbb{Z}_pG \simeq \mathbb{Z}_pH;$$

because in that case, in view of (5.8) and (5.9), one would have obtained just as much information in the modular case as in the integral case.

In the integral case we have proved

(5.12) G finite, nilpotent of class two, $\mathbb{Z}G \simeq \mathbb{Z}H \Rightarrow G \simeq H$.

The group G is nilpotent of class two $\Leftrightarrow \gamma_3(G) = 1 \Leftrightarrow D_{3,\mathbb{Z}}(G) = 1$ So the modular analogue of (5.12) is

(5.13) $\mathbb{Z}/p\mathbb{Z}G \simeq \mathbb{Z}/p\mathbb{Z}H$, $D_{3,\mathbb{Z}/p\mathbb{Z}}(G) = 1 \Rightarrow G \simeq H$.

This turns out to be true and we look into this in detail in the next section.

6. MODULAR GROUP ALGEBRAS

We shall first prove that the center $\zeta(G)$ of a finite p-group G is an invariant in the modular group algebra FG. We need the

Proposition 6.1. Let G be a finite Abelian group and F a field of characteristic $p > 0$. Let

$$G_i = \{g \in G: g^{p^i} = 1\} = \{g_1, \cdots, g_\ell\}, \quad \Gamma_i = \{\gamma \in FG: \gamma^{p^i} = 0\}.$$

Then (i) $\Gamma_i = \Delta(G, G_i)$ and (ii) Γ_i is a vector space over F of dimension $(G:1) - (G:G_i)$.

Proof: Clearly Γ_i is an F-space. Let T, containing 1, be a transversal for G_i in G. Then any element $\gamma \in FG$ can be written as

$$\gamma = \sum_k t_k \sum_{j=1}^{\ell} a_{kj} g_j, \quad a_{kj} \in F, \quad t_k \in T.$$

Then

$$\gamma^{p^i} = \sum_k t_k^{p^i} \sum_{j=1}^{\ell} a_{kj}^{p^i} g_j^{p^i} = \sum_k \left(\sum_{j=1}^{\ell} a_{kj}^{p^i} \right) t_k^{p^i}.$$

So,

$$\gamma^{p^i} = 0 \iff \left(\sum_{j=1}^{\ell} a_{kj}^{p^i} \right) = \left(\sum_{j=1}^{\ell} a_{kj} \right)^{p^i} = 0 \quad \text{for all } k$$

$$\iff \sum_{j=1}^{\ell} a_{kj} = 0 \quad \text{for all } k$$

$$\iff \gamma \in \Delta(G, G_i).$$

We have proved that $\Gamma_i = \Delta(G,G_i)$. But $FG/\Delta(G,G_i) \simeq F(G/G_i)$, the latter having F-dimension $(G:G_i)$. Hence Γ_i has F-dimension $(G:1) - (G:G_i)$ as stated. □

Since the set of numbers $|G_i|$ characterizes an Abelian p-group G we have the following corollaries. The first one is a special case of (3.2).

Corollary 6.2 (Deskins [1]). Let G be a finite Abelian p-group and F a field of characteristic p. Then

$$FG \simeq_F FH \Rightarrow G \simeq H.$$

Corollary 6.3. Let G be an Abelian group. Suppose for every prime p dividing $|G|$, there exists a field F_p of characteristic p such that $F_pG \simeq_{F_p} F_pH$. Then $G \simeq H$.

Lemma 6.4. The center $\zeta(RG)$ of the group ring of a group G over a commutative ring R is the R-span of finite class sums of G i.e. $\zeta(RG) = \{\Sigma a_x C_x : C_x = \Sigma_{y\sim x} y,\ a_x \in R\}$.

Proof: Let $\gamma = \Sigma\gamma(g)g \in RG$. Then for $x \in G$,

$$\gamma^x = x^{-1}\gamma x = \sum\gamma(g)g^x.$$

Therefore, $\gamma^x = \gamma \iff \gamma(g) = \gamma(g^x)$. Hence, $\gamma \in \zeta(FG) \iff \gamma(g) = \gamma(g^x)$ for all $x \in G$, $g \in G$. This proves the result. □

Lemma 6.5. Let $RG^{[2]}$ denote the R-span of $\{[\gamma,\mu] = \gamma\mu - \mu\gamma,\ \gamma,\mu \in RG\}$. Then

$$RG^{[2]} = \{\alpha = \sum\alpha(g)g \in RG : \tilde{\alpha}(g) = \sum_{x\sim g} \alpha(x) = 0 \quad \text{for all} \quad g \in G\}.$$

Proof: First suppose $\tilde{\alpha}(g) = 0$ for all $g \in G$. Then α is a linear combination of terms of the form $(g - x^{-1}gx)$, $g \in G$, $x \in G$. So $\alpha = \Sigma r[x^{-1}, gx]$, $r \in R$ and $\alpha \in RG^{[2]}$. The reverse containment is proved in (I.1.5). □

Theorem 6.6 (Ward [1], Sehgal [1]). Let G be a finite p-group and F a field of characteristic p. Then

$$FG \simeq_F FH \Rightarrow \zeta(G) \simeq \zeta(H).$$

Proof: Since the F-dimension of FG = F-dimension of $FH = |G|$, H is also a p-group. Due to (6.4), (6.5) and the fact that each conjugacy class of G has a p-power number of elements it follows that

$$FG^{[2]} \cap \zeta(FG) = \left\{\sum a_x C_x : C_x = \sum_{y\sim x} y,\ a_x \in F,\quad x \notin \zeta(G)\right\}.$$

Again, using the fact that $g_1g_2 = z \in \zeta(G) \Rightarrow g_1^x g_2^x = z$ for $x \in G$, we conclude that $FG^{[2]} \cap \zeta(FG)$ is an ideal in $\zeta(FG)$. Thus

$$\zeta(FG)/[FG^{[2]} \cap \zeta(FG)] \simeq F(\zeta(G)).$$

Since $\zeta(FG)$ and $FG^{[2]}$ are mapped under the isomorphism $FG \simeq FH$ onto their counterparts in FH we conclude that $F\zeta(G) \simeq F\zeta(H)$. Thus by (6.2), $\zeta(G) \simeq \zeta(H)$. □

Let us recall (1.22) that the Brauer-Jennings-Zassenhaus M-series of a group G at a prime p is characterized by

$$(6.7)\quad M_{n,p}(G) = D_{n,\mathbb{Z}/p\mathbb{Z}}(G) = \left\{g \in G:\ g-1 \in \Delta^{n}_{\mathbb{Z}/p\mathbb{Z}}(G)\right\} = \prod_{ip^j \geq n} \gamma_i(G)^{p^j}$$

where $\gamma_i(G)$ denotes the i-th term in the lower central series of G. We shall prove that for each $n \geq 1$, the factors $M_{n,p}/M_{n+1,p}$ and $M_{n,p}/M_{n+2,p}$ are isomorphism invariants of $\mathbb{Z}/p\mathbb{Z}G$. In the following since p will be a fixed prime we shall omit the subscript p from $M_{n,p}(G)$ and denote it by $M_n(G)$. We shall also omit the subscript R from Δ_R when no confusion can arise.

The Lie central series $\{\gamma_i R\} = \{R^{[i]}\}$ of a ring R is defined inductively by $R^{[1]} = R$, $R^{[i+1]} = [R^{[i]},R]$, the additive group generated by all $[\alpha,\beta] = \alpha\beta - \beta\alpha$, $\alpha \in R^{[i]}$, $\beta \in R$, for all $i \geq 1$.

<u>Lemma 6.8</u>. $\Delta_R^{[n]}(G) + \Delta_R^{n+1}(G) = \Delta_R\big(G,\gamma_n(G)\big) + \Delta_R^{n+1}(G) = \Delta_R\big(\gamma_n(G)\big) + \Delta_R^{n+1}(G)$ for all $n \geq 1$.

<u>Proof:</u> The last equality follows from the fact

$$(6.9)\qquad g \in \gamma_i(G) \Rightarrow g-1 \in \Delta^i(G)$$

and the equation

$$(6.10)\qquad x(g-1) = g-1 + (x-1)(g-1).$$

In order to prove that the first two expressions are equal, we use induction on n. For $n = 1$, the result is trivial as both sides equal $\Delta(G)$. Let us assume

(6.11) $\Delta^{[i]}(G) + \Delta^{i+1}(G) = \Delta(G,\gamma_i(G)) + \Delta^{i+1}(G)$.

Let $[x,y]$, $x \in \Delta(G)$, $y \in \Delta^{[i]}(G)$, be a generator of $\Delta^{[i+1]}(G)$. Then by (6.11)

$$\begin{aligned} [x,y] &= [x,y_1+y_2], \quad y_1 \in \Delta(\gamma_i(G)), \quad y_2 \in \Delta^{i+1}(G) \\ &\equiv [x,y_1] \pmod{\Delta^{i+2}(G)} \\ &\equiv \sum_{g\in\gamma_i(G)} a_g[x,(g-1)] \pmod{\Delta^{i+2}(G)}, \quad a_g \in R. \end{aligned}$$

Writing $x = \Sigma_{h\in G} b_h(h-1)$, $b_h \in R$ we have

$$\begin{aligned} [x,y] &\equiv \sum_{g,h} a_g b_h [h,g] \pmod{\Delta^{i+2}(G)}, \quad g \in \gamma_i(G), \quad h \in G, \\ &\equiv \sum_{g,h} a_g b_h hg(1-g^{-1}h^{-1}gh) \pmod{\Delta^{i+2}(G)}. \end{aligned}$$

But since $1 - g^{-1}h^{-1}gh \in \gamma_{i+1}(G)$ we have proved that

$$\Delta^{[i+1]}(G) + \Delta^{i+2}(G) \subseteq \Delta(G,\gamma_{i+1}(G)) + \Delta^{i+2}(G).$$

In view of the equation (6.10), in order to prove (6.11) it suffices to prove

(6.12) $g \in \gamma_{i+1}(G) \Rightarrow g-1 \equiv \alpha \bmod \Delta^{i+2}(G)$ for some $\alpha \in \Delta^{[i+1]}(G)$.

Because of the equation

$$uv-1 = (u-1) + (v-1) + (u-1)(v-1),$$

we may assume that $g = xg_1x^{-1}g_1^{-1}$, $g_1 \in \gamma_i(G)$, $x \in G$. Then

$$\begin{aligned} g-1 &= xg_1x^{-1}g_1^{-1} - 1 = (xg_1-g_1x)x^{-1}g_1^{-1} \\ &= (xg_1-g_1x)(x_1^{-1}g_1^{-1}-1) + (xg_1-g_1x) \end{aligned}$$

$$\equiv xg_1 - g_1x \pmod{\Delta^{i+2}(G)}$$

$$\equiv [x-1, g_1 - 1] \pmod{\Delta^{i+2}(G)}$$

$$\equiv [x-1, \alpha_1+\alpha_2] \pmod{\Delta^{i+2}(G)}, \ \alpha_1 \in \Delta^{[i]}(G), \ \alpha_2 \in \Delta^{i+1}(G)$$

(by (6.11))

$$\equiv [x-1, \alpha_1] \pmod{\Delta^{i+2}(G)}.$$

Since $[x-1,\alpha_1] \in \Delta^{[i+1]}(G)$, the lemma is proved. □

The next theorem was proved for finite p-groups by Dieckman [1] and Hill [1] and in general by Passi and Sehgal [1]. See also Quillen [1].

Theorem 6.13. Let F be the field of p elements. Then

$$FG \simeq FH \Longrightarrow M_n(G)/M_{n+1}(G) \simeq M_n(H)/M_{n+1}(H) \quad \text{for all } n \geq 1.$$

Proof: Define

$$I_n(G) = \sum_{ip^j \geq n} \left(\Delta^{[i]}(G)\right)^{\underline{p^j}} + \Delta^{n+1}(G)$$

where by $S^{\underline{p^j}}$ for a subset S of FG we understand the additive subgroup of FG generated by all s^{p^j}, $s \in S$. We assert that

(6.14) $\quad m \in M_n(G) \Longrightarrow m-1 \in I_n(G).$

Let $g \in \gamma_i(G)$ and $ip^j \geq n$ so that $g^{p^j} \in M_n(G)$ then by (6.8) we have

$$g-1 \equiv \alpha \bmod \Delta^{i+1}(G), \quad \alpha \in \Delta^{[i]}(G).$$

Then

$$g^{p^j}-1 \equiv \alpha^{p^j} \bmod \Delta^{n+1}(G)$$

and $g^{p^j}-1 \in I_n(G)$. Now (6.14) follows easily. Therefore we can define a map

$$\theta\colon M_n(G) \to I_n(G)/\Delta^{n+1}(G)$$

by $\theta(m) = \overline{m-1}$. We claim that

(6.15) θ is an epimorphism.

Let $\alpha \in \Delta^{[i]}(G)$ and $ip^j \geq n$. Then by (6.8)

$$\alpha = \sum_{g\in\gamma_i(G)} a_g(g-1) + \delta, \quad a_g \in F, \quad \delta \in \Delta^{i+1}(G).$$

We have, therefore,

$$\alpha \equiv (h-1) + \delta_1, \quad h \in \gamma_i(G), \quad \delta_1 \in \Delta^{i+1}(G).$$

Consequently,

$$\alpha^{p^j} \equiv (h^{p^j}-1) \bmod \Delta^{n+1}(G)$$

which implies that $\theta(h^{p^j}) = \alpha^{p^j}$ and θ is an epimorphism. Since the kernel of θ is $M_{n+1}(G)$ we have the isomorphism

(6.16) $M_n(G)/M_{n+1}(G) \simeq I_n(G)/\Delta^{n+1}(G)$.

Since $\Delta(G)$ and $I_n(G)$ are mapped under the isomorphism $FG \simeq FH$ to $\Delta(H)$ and $I_n(H)$ respectively we have proved that

$$M_n(G)/M_{n+1}(G) \simeq M_n(H)/M_{n+1}(H). \quad \square$$

One can further prove that the quotients $M_n(G)/M_{n+2}(G)$ are also isomorphism invariants of the group algebra $\mathbb{Z}/p\mathbb{Z}G$.

Theorem 6.17 (Passi and Sehgal [1]). Let F be the field of p elements. Then

$$FG \simeq FH \Rightarrow M_n(G)/M_{n+2}(G) \simeq M_n(H)/M_{n+2}(H) \quad \text{for all } n \geq 1.$$

Proof: Consider the embedding

$$M_n(G)/M_{n+1}(G) \to \Delta^n(G)/\Delta^{n+1}(G)$$

given by $\overline{m} \to \overline{m-1}$ for $m \in M_n(G)$. Since $\Delta^n(G)/\Delta^{n+1}(G)$ is a vector space over F we have

$$(6.18) \quad \Delta^n(G)/\Delta^{n+1}(G) = \left[\Delta\left(M_n(G)\right) + \Delta^{n+1}(G)\right]/\Delta^{n+1}(G) \oplus K_n(G)/\Delta^{n+1}(G)$$

where $K_n(G)$ is a subspace of FG. Since $\Delta^{n+1}(G) \subseteq K_n(G) \subseteq \Delta^n(G)$,

$$g \in G, \quad k \in K_n(G) \Longrightarrow gk = (g-1)k + k \in K_n(G).$$

Thus $K_n(G)$ is an ideal of FG. We claim that

$$(6.19) \quad M_n(G)/M_{n+2}(G) \simeq [M_n(G) + K_{n+1}(G)]/K_{n+1}(G) \subseteq \mathcal{U}\left(FG/K_{n+1}(G)\right).$$

To see this, define the natural map

$$\lambda\colon M_n(G) \to [M_n(G) + K_{n+1}(G)]/K_{n+1}(G), \quad \lambda(m) = \overline{m}, \quad m \in M_n(G).$$

Clearly λ is an epimorphism containing $M_{n+2}(G)$ in its kernel. Suppose $\lambda(m) = \overline{1}$ i.e. $m-1 \in K_{n+1}(G)$. Then $m-1 \in \Delta^{n+1}(G)$ and $m \in M_{n+1}(G)$. Thus by (6.18),

$$m-1 \in \Delta\left(M_{n+1}(G)\right) \cap K_{n+1}(G) = \Delta^{n+2}(G)$$

and $m \in M_{n+2}(G)$ proving (6.19). Let $\theta\colon FG \to FH$ be the given isomorphism. Then we claim that

$$(6.20) \qquad \theta\left(\Delta(M_n(G)) + \Delta^{n+1}(G)\right) = \Delta\left(M_n(H)\right) + \Delta^{n+1}(H),$$

$$(6.21) \qquad \theta\left(M_n(G) + K_{n+1}(G)\right) = M_n(H) + \theta\left(K_{n+1}(G)\right).$$

We complete the proof of the theorem assuming (6.20) and (6.21). Applying θ to (6.18) and using (6.20) we get

$$(6.22) \quad \Delta^n(H)/\Delta^{n+1}(H) = \left[\Delta\left(M_n(H)\right)+\Delta^{n+1}(H)\right]/\Delta^{n+1}(H) \oplus \theta\left(K_n(G)\right)/\Delta^{n+1}(H).$$

This splitting gives the counterpart of (6.19), namely

$$\begin{aligned} M_n(H)/M_{n+2}(H) &\simeq \Big[M_n(H) + \theta\big(K_{n+1}(G)\big)\Big]/\theta\big(K_{n+1}(G)\big) \\ &\simeq \Big[M_n(G) + K_{n+1}(G)\Big]/K_{n+1}(G) \quad \text{(by (6.21))} \\ &\simeq M_n(G)/M_{n+2}(G) \qquad\qquad \text{(by (6.19)).} \end{aligned}$$

It remains only to prove (6.20) and (6.21) which we do now.

6.23. Proof of (6.20)

Let $x \in \Delta\big(M_n(G)\big) + \Delta^{n+1}(G)$; then x can be written as $x = m-1 + \alpha$, $m \in M_n(G)$, $\alpha \in \Delta^{n+1}(G)$. But by (6.14), $m-1 \in I_n(G)$. Therefore, $\theta(x) = \theta(m-1) + \theta(\alpha) \in I_n(H)$. Again by (6.15) the map $y \to \overline{y-1}$ is an epimorphism $M_n(H) \to I_n(H)/\Delta^{n+1}(H)$ and thus $\theta(x) = h-1 + \gamma$, $h \in M_n(H)$, $\gamma \in \Delta^{n+1}(H)$, proving (6.20). □

6.24. Proof of (6.21)

Let $m \in M_n(G)$. Then $\theta(m-1)$ being an element of $I_n(H)$ it follows by (6.15) that $\theta(m-1) = h-1+\gamma$, $h \in M_n(H)$, $\gamma \in \Delta^{n+1}(H)$. Due to the splitting of (6.22) we have

$$\gamma = u - 1 + v, \quad u \in M_{n+1}(H), \quad v \in \theta\big(K_{n+1}(G)\big).$$

Therefore,

$$\begin{aligned} \theta(m-1) &= h - 1 + u - 1 + v \\ &= hu - 1 + w, \quad w = v - (h-1)(u-1). \end{aligned}$$

As $(h-1)(u-1) \in \Delta^{n+2}(H) = \theta\big(\Delta^{n+2}(G)\big) \subseteq \theta\big(K_{n+1}(G)\big)$, $w \in \theta\big(K_{n+1}(G)\big)$. Thus

$$\theta\big(M_n(G) + K_{n+1}(G)\big) \subseteq M_n(H) + \theta\big(K_{n+1}(G)\big).$$

Now in view of (6.20), $\theta\big(K_{n+1}(G)\big)$ acts as $K_{n+1}(H)$ in (6.18) and we conclude

$$\theta^{-1}\Big(M_n(H) + \theta\big(K_{n+1}(G)\big)\Big) \subseteq M_n(G) + K_{n+1}(G).$$

This proves (6.21). □

Corollary 6.25. If G and H are groups of M-length ≤ 2 i.e. $M_{3,p}(G) = M_{3,p}(H) = 1$ for a prime p then $\mathbb{Z}/p\mathbb{Z}G \simeq \mathbb{Z}/p\mathbb{Z}H \Rightarrow G \simeq H$.

Actually it is known (see Passi and Sehgal [1]) that $M_{n,p}(G) = 1 \Leftrightarrow M_{n,p}(H) = 1$.

7. AUTOMORPHISMS OF $\mathbb{Z}G$

The following result regarding the automorphisms of a central simple algebra is well known.

Theorem 7.1 (Noether-Skolem). Any automorphism of a finite dimensional simple algebra S over a field F, keeping its centre $\zeta(S)$ fixed elementwise, is inner.

Proof: See Herstein [2].

We shall use this theorem to prove a similar result regarding the automorphisms of $\mathbb{Z}G$ for a small class of groups G. We recall

Theorem 4.1. Let $\theta: \mathbb{Z}G \to \mathbb{Z}H$ be a normalized isomorphism where G is a finite group. Then for any class sum K_x, $x \in G$, of G, we have $\theta(K_x) = K_y$, $y \in H$, a class sum of H.

Proposition 7.2. Let θ be a normalized automorphism of the integral group ring $\mathbb{Z}G$ of a finite group G. Let C_i, $1 \le i \le r$, be the conjugacy classes and K_i the corresponding class sums of G. Suppose that $\theta(K_i) = K_{i'}$, $1 \le i,\ i' \le r$ and that there exists an automorphism σ of G such that $\sigma(C_i) = C_{i'}$ for all $1 \le i \le r$. Then we can find a unit $\gamma \in QG$ such that

$$\theta(g) = \gamma g^{\sigma} \gamma^{-1} \quad \text{for all } g \in G.$$

Proof: Extend σ and θ to QG. By (6.4) the center of QG is generated by the class sums K_i, $1 \le i \le r$ and thus is kept elementwise fixed by $\sigma^{-1}\theta$. Since QG is semisimple, we can write

$$QG = S_1 \oplus \cdots \oplus S_t,$$

a direct sum of simple rings S_i. Let $1 = e_1 + \cdots + e_t$, where $e_i \in S_i$ are central idempotents. Then

$$S_i = e_i QG, \quad (\sigma^{-1}\theta)S_i = S_i, \quad 1 \le i \le t.$$

As $\sigma^{-1}\theta$ keeps the center of S_i fixed elementwise, it acts on S_i as an inner automorphism by some $\alpha_i \in S_i$ (7.1). It follows that $\sigma^{-1}\theta$ is an inner automorphism of QG by $\alpha = \alpha_1 + \cdots + \alpha_t$. Thus we have for $g \in G$, that

$$(\sigma^{-1}\theta)(g) = \alpha g \alpha^{-1}, \quad \theta(g) = \gamma g^{\sigma} \gamma^{-1} \quad \text{where} \quad \gamma = \sigma(\alpha).$$

Proposition 7.3. Let θ be a normalized automorphism of $\mathbb{Z}G$, where G is a finite nilpotent group of class two. Let K_i, $1 \le i \le r$, be the class sums of G. Suppose that $\theta(K_i) = K_{i'}$, $1 \le i \le r$. Then there exists an automorphism σ of G which when extended to $\mathbb{Z}G$, satisfies $\sigma(K_i) = K_{i'}$ for all $1 \le i \le r$.

Proof: We first observe that a class sum of G is of the form $g\hat{H}$ where $\hat{H} = \Sigma_{h \in H} h$ and H is a subgroup of the derived group G'. Let $\theta(g) = \gamma$ and let K_g be the class sum of g. Then $\theta(K_g) = \theta(g\hat{H}) = \gamma\theta(\hat{H}) = \gamma\hat{H}_1$ for another subgroup H_1, by using (5.2). Now $\gamma\hat{H}_1 = K_{g_1} = g_1\hat{H}_2$ for a subgroup H_2. We claim that $H_1 = H_2$. Observe that

$$|H_1| g_1\hat{H}_2 = |H_1|\gamma\hat{H}_1 = \gamma\hat{H}_1\hat{H}_1 = g_1\hat{H}_2\hat{H}_1$$

and thus

$$|H_1|\hat{H}_2 = \hat{H}_2\hat{H}_1 .$$

Therefore H_1 is contained in H_2 and by symmetry $H_1 = H_2$. Thus we have

$$\theta(K_g) = \gamma\hat{H}_1 = g_1\hat{H}_1.$$

Next we claim that there exists a $g_\gamma \in G$ such that $\gamma \equiv g_\gamma$ mod $\Delta(G,H_1)\Delta(G)$. Since $\gamma\hat{H}_1 = g_1\hat{H}_1$ we have by (4.18), $(\gamma - g_1) \in \Delta(G,H_1)$. Thus

$$\begin{aligned}
\gamma &= g_1 + \sum_{h\in H_1} (h-1)t(h), \quad t(h) \in \mathbb{Z}G \\
&\equiv g_1 + \sum_h (h-1)n_h \ \big(\mathrm{mod}\ \Delta(G,H_1)\Delta(G)\big), \quad n_h \in \mathbb{Z} \\
&\equiv g_1 + \Big(\prod_h h^{n_h} - 1\Big) \big(\mathrm{mod}\ \Delta(G,H_1)\Delta(G)\big) \\
&\equiv g_1 \prod_h h^{n_h} \big(\mathrm{mod}\ \Delta(G,H_1)\Delta(G)\big) \\
&\equiv g_\gamma \ \big(\mathrm{mod}\ \Delta(G,H_1)\Delta(G)\big).
\end{aligned}$$

Now we define a map $\sigma: G \to G$ by $\sigma(g) = g_\gamma$. That this map is a well defined automorphism of G is proved in (5.3) in view of the fact that

$$\gamma \equiv g_\gamma \ \mathrm{mod}\big(\Delta(G,G')\Delta(G)\big).$$

Moreover,

$$\sigma(K_g) = \sigma(g\hat{H}) = \sigma(g)\sigma(\hat{H}) = g_\gamma\theta(\hat{H}) = g_\gamma\hat{H}_1 = \gamma\hat{H}_1 = \theta(K_g),$$

completing the proof of the proposition. □

<u>Theorem 7.4</u> (Sehgal [2]). Let θ be a normalized automorphism of $\mathbb{Z}G$, where G is a finite nilpotent group of class two. Then there exists an automorphism σ of G and a unit γ of QG such that $\theta(g) = \gamma g^\sigma \gamma^{-1}$, for all $g \in G$.

<u>Proof</u>: Theorem 4.1, Proposition 7.2 and Proposition 7.3.

Based on this small evidence we make the

Conjecture 7.5. Let θ be a normalized automorphism of $\mathbb{Z}G$. Then there exists an automorphism σ of G and a unit γ of QG such that $\theta(g) = \gamma g^{\sigma}\gamma^{-1}$, for all $g \in G$.

This conjecture has been confirmed by G. Peterson ([1],[2]) for the symmetric groups S_n and certain classes of finite metabelian groups.

CHAPTER IV

UNIQUENESS OF THE COEFFICIENT RING IN RG

In the last chapter we asked if G is an invariant of the group ring RG. We may, just as naturally, ask whether or not R is an invariant of RG. More precisely; if for two rings R and S and a fixed group G, RG is isomorphic to SG as a ring, does it follow that R is isomorphic to S? The answer is 'no' in general. For example, let $G_0, G_1, G_2, \cdots$ be infinite cyclic groups and

$$R = \mathbb{Z}G_0, \quad S = \mathbb{Z} \quad \text{and} \quad G = \prod_{i \geq 1} G_i \ ,$$

the direct product of $G_1, G_2, \cdots$. Then

$$RG = (\mathbb{Z}G_0)G \simeq \mathbb{Z}(G_0 \times G) \simeq \mathbb{Z}G = SG$$

but certainly $\mathbb{Z}G_0$ is not isomorphic to $\mathbb{Z}$. This happens because G is too large. In analogy with what we did in the last chapter we restrict G to be an infinite cyclic group $\langle x \rangle$ and pose the following

<u>Problem</u>: Does $R\langle x\rangle \simeq S\langle x\rangle$ imply $R \simeq S$?

The results are fragmentary; there are just a few classes of rings for which an affirmative answer is known. We do not know any counterexamples. Also, it is known that if R and S are commutative then $R\langle x\rangle \simeq S\langle x\rangle$ implies that R and S are subisomorphic. Here, two rings are said to be subisomorphic if each is isomorphic to a subring of the other. Throughout this chapter, $\langle x\rangle$ will denote an infinite cyclic group and of course R will have a 1. A problem similar to the above problem, namely, whether or not the isomorphism of polynomial rings R[x] and S[X] implies $R \simeq S$, has been considered by Coleman and Enochs [1], Abhyanker, Heinzer and Eakin [1] and some others. An example of two non-isomorphic commutative rings with isomorphic polynomial rings was given by Hochster [1].

1. AUTOMORPHISMS OF $R\langle x\rangle$

The following is an extension of (II.3.3) which we restate here as we shall need it.

<u>Lemma 1.0</u>. Suppose that R is a commutative ring with 0 and 1 as its only idempotents. If R has no nonzero nilpotent elements then

$$\mathcal{U}(R\langle x\rangle) = \mathcal{U}(R) \cdot \langle x\rangle \ .$$

<u>Lemma 1.1</u> (Parmenter). Let $u = \Sigma a_i x^i \in R\langle x\rangle$ where R is a commutative ring. Then u is a unit if and only if there exist $b_j \in R$ satisfying $\Sigma a_i b_{-i} = 1$ and $a_i b_j$ nilpotent for $j \neq -i$.

<u>Proof</u>: Suppose that $uv = 1$ for $v = \Sigma b_j x^j$. Let P be a prime ideal of R. Then since units of $R/P\langle x\rangle$ are trivial we have

$$u \equiv a_r x^r, \quad v \equiv b_{-r} x^{-r} \text{ mod } P\langle x\rangle, \quad \text{for some } r.$$

This implies that $a_i a_j, b_i b_j \in P$ for all $i \neq j$ and $a_s b_t \in P$ for all $s \neq -t$. Since P is arbitrary and $\bigcap_P P$ is nilpotent we can conclude that $a_i a_j, a_s a_t, b_i b_j$ are nilpotent for $i \neq j$ and $s \neq -t$. Also, comparing the coefficient of identity in the equation $uv = 1$ we see that $1 = \Sigma a_i b_{-i}$.

Conversely, suppose that a_i and b_j satisfy the given conditions. Then

$$u\left(\sum_j b_j x^j\right) = 1 - \alpha$$

where α is nilpotent with say $\alpha^\ell = 0$. Thus

$$u\left(\sum_j b_j x^j\right)(1+\alpha+\cdots+\alpha^{\ell-1}) = 1$$

and u is a unit of $R\langle x\rangle$. $\square$

The above argument gives the following useful fact which we record for use later on.

Lemma 1.2. If $u = \Sigma a_i x^i \in R\langle x\rangle$ has inverse $v = \Sigma b_j x^j$ and R is commutative then $\Sigma_i a_i b_{-i} = 1$ and $a_i a_j$, $a_s b_t$, $b_i b_j$ are nilpotent for $i \neq j$ and $s \neq -t$.

Lemma 1.3 (Parmenter). Let R be commutative and $u = \Sigma a_i x^i$ be a unit of $R\langle x\rangle$ with a_0 nilpotent (or zero). Then $\{u^j : j \in \mathbb{Z}\}$ is a linearly independent set over R.

Proof: Let P be a prime ideal of R. Write

$$\bar{u} = \Sigma \bar{a}_i x^i \in (R/P)\langle x\rangle.$$

Since a_0 is nilpotent it belongs to all prime ideals of R and thus $\bar{a}_0 = 0$. Because $\bar{u}$ is a unit at least one $\bar{a}_i \neq 0$. Let m be the largest i such that $\bar{a}_i \neq 0$. We have

$$\bar{u} = \sum_{\ell}^{m} \bar{a}_i x^i \qquad \ell \le m.$$

Suppose that $\Sigma c_j u^j = 0$ for $c_j \in R$. We may suppose that all $j > 0$. Let t be the largest j with $c_j \neq 0$. Then consider the equation

$$\text{(i)}\quad \sum_j c_j \left(\sum_i a_i x^i\right)^j = 0$$

mod P. The coefficient of x^{mt}, $c_t a_m^t$, belongs to P. But since $a_m \notin P$ it follows that c_t and thus all c_i belong to P. Since P is an arbitrary prime ideal it follows that c_i is nilpotent for all i.

Let us recall (1.2) that if $u^{-1} = \Sigma b_\ell x_\ell$ then

(ii) $\sum a_i b_{-i} = 1$, $a_i a_j, b_i b_j, a_r b_s$ nilpotent for $i \neq j$, $r \neq -s$.

We have seen that (i) implies c_i is nilpotent for all i. Let I be the ideal of R generated by $\{c_i\} \cup \{a_i : a_i \text{ nilpotent}\} \cup \{a_i a_j : i \neq j\}$. Suppose that all $c_i \in I^k$ for some $k \geq 1$. We intend to show that $c_i \in I^{k+1}$. At first we claim

(iii) $c_i a_j \in I^{k+1}$ for all i,j.

Since a_0 is nilpotent $c_i a_0$ does belong to I^{k+1} for all i. The equation (i) implies by (ii) that

$$\sum_j c_j \left(\sum_i a_i^j x^{ij} \right) \equiv 0 \pmod{I^{k+1}<x>}.$$

Multiplying by a fixed a_r, $r \neq 0$, we conclude

$$\sum_j c_j a_r^{j+1} x^{rj} \equiv 0 \pmod{I^{k+1}<x>}.$$

It follows that $c_j a_r^{j+1} \in I^{k+1}$. Thus by (ii) we have

$$c_j a_r^j = \sum c_j a_r^j a_i b_{-i} \equiv c_j a_r^{j+1} b_{-r} \equiv 0 \pmod{I^{k+1}}.$$

Repeating this we conclude that $c_j a_r \in I^{k+1}$ for all j, $r \neq 0$ as claimed in (iii). Now multiplying the relation $1 = \Sigma a_i b_{-i}$ by c_j we conclude that $c_j \in I^{k+1}$.

Since I is nilpotent it follows that $c_j = 0$ for all j. □

Lemma 1.4 (Parmenter). Let R be a commutative ring. Let $u = \Sigma_i a_i x^i$ be a unit of R<x>. Then $x \to u$ induces an R-endomorphism of R<x>. This is an automorphism if and only if a_i is nilpotent for all $i \neq \pm 1$.

Proof: Suppose that $x \to \Sigma a_i x^i$ is an R-automorphism. We wish to prove that a_i is nilpotent for $i \neq \pm 1$. After factoring by N the ideal of nilpotent elements of R, we may assume that R has no nilpotent elements and prove that a_i is zero for $i \neq -1,1$. Let

$$\left(\sum a_i x^i\right)^{-1} = \sum b_j x^j .$$

Then by (1.2) we have $a_i a_j = 0 = b_i b_j$ for $i \neq j$ and $a_s b_t = 0$ for $s + t \neq 0$. Since the map is onto there exist elements $c_j \in R$ with

$$x = \sum c_j\left(\sum a_i x_i\right)^j$$

$$= \sum_{j \geq 0} c_j\left(\sum a_i x^i\right)^j + \sum_{j>0} c_{-j}\left(\sum b_i x^i\right)^j .$$

Comparing the coefficient of x we have

$$1 = c_1 a_1 + c_{-1} b_1$$

because all $a_i a_j = 0 = b_i b_j$ for $i \neq j$. Multiplying by a_j, $j \neq 1,-1$ we get $a_j = 0$.

Conversely, assume that $\Sigma a_i x^i$ is a unit with a_j nilpotent for $j \neq 1,-1$. We wish to prove that $x \to \Sigma a_i x^i$ induces an R-automorphism of R<x>. We have proved in (1.3) that this map is 1-1. To prove that it is, indeed, an epimorphism, write

$$u = \sum a_i x^i, \quad u^{-1} = \sum b_j x^j .$$

Then

$$\sum a_i b_{-i} = 1, \; a_i a_j, b_i b_j, a_s b_t \text{ are nilpotent for } i \neq j,\; s+t \neq 0. \quad (*)$$

Observe that

$$b_{-1}u + a_{-1}u^{-1} = (b_{-1}a_1+a_{-1}b_1)x + \sum_{i\neq 1}(b_{-1}a_i+a_{-1}b_i)x^i$$

$$= x + n_1, \quad n_1 \in I\langle x\rangle$$

where I is the nilpotent ideal generated by $\{a_ib_j, \ i\neq -j\} \cup \{a_i : i\neq 1,-1\}$. Also,

$$(b_{-1}u+a_{-1}u^{-1})(a_1u^{-1}+b_1u) = (a_1b_{-1}+a_{-1}b_1) + b_{-1}b_1u^2 + a_{-1}a_1u^{-2}$$

$$= 1 + \beta,$$

where β is nilpotent. Thus $(x+n_1)^{-1} \in R\langle u\rangle$. We assert that $x + n_1$ and $(x+n_1)^{-1}$ generate $R\langle x\rangle$ over R. Let us write $n_1 = \Sigma_j d_i x^i$ and use induction on the index of nilpotency of I. Clearly,

$$x + n_1 - \sum_i d_i(x+n_1)^i = x + n_2$$

where $n_2 \in I^2\langle x\rangle$. Thus by induction, $x + n_2$ and $(x+n_2)^{-1}$ together generate $R\langle x\rangle$ over R. The assertion now follows easily from the last equation and the map is onto. □

2. JACOBSON RADICAL OF $R\langle x\rangle$

It has been proved by Amitsur [1] that the Jacobson radical, $J(R[x])$, of the polynomial ring $R[x]$, over an arbitrary ring R with 1, is given by $N[x]$ where $N = J(R[x]) \cap JR$ is a nil ideal. We do not know if a corresponding result holds in $R\langle x\rangle$. However, we do have the following result, the polynomial analogue of which is also proved in Herstein [2].

Proposition 2.1. If R has no nil ideals then $J(R\langle x\rangle) = 0$.

Proof: Suppose that $J(R\langle x\rangle) \neq 0$. Pick

$$r = a_0 x^{n_0} + a_1 x^{n_1} + \cdots + a_k x^{n_k}, \quad n_0 < n_1 < \cdots < n_k$$

a nonzero element of $J(R\langle x\rangle)$ with as few nonzero coefficients a_i as possible. We may also suppose that $n_0 > 0$. Now, $a_i r - r a_i$ is also an element of $J(R\langle x\rangle)$ and has a smaller number of nonzero coefficients. Thus $a_i r - r a_i = 0$ and $a_i a_j = a_j a_i$ for all i and j.

Since $rx \in J(R\langle x\rangle)$ we have for some $s \in J(R\langle x\rangle)$, $rx + s + srx = 0$. Consequently, $s = -rx - srx = -rx - (-rx-srx)rx = -rx + r^2x^2 + sr^2x^2$. Continuing in this manner we obtain

$$s = -rx + r^2x^2 - r^3x^3 + \cdots + (-1)^n r^n x^n + (-1)^n s r^n x^n.$$

Picking n large enough we conclude that $r,s \in R_0\langle x\rangle$ where R_0 is the commutative ring generated by $1, a_0, \cdots, a_k$. We have

$$(1+s)(1+rx) = 1, \quad rx = \sum_0^k a_i x^{n_i+1}, \quad s = \sum s_j x^j .$$

It follows by (1.2) that $1a_i = a_i$ is nilpotent for all i. Now, let

$$I = \{a \in R\colon ax^{n_0} + b_1 x^{n_1} + \cdots + b_k x^{n_k} \in J(R\langle x\rangle) \text{ for some } b_1, \cdots, b_k \in R\}.$$

Then I is an ideal of R containing a_0. Moreover, as we have seen, I is nil which is a contradiction proving our result. □

Corollary 2.2. $J(R\langle x\rangle) \subseteq (JR)\langle x\rangle$.

Proof: Since R/JR has no nil ideals

$$J\big((R/JR)\langle x\rangle\big) = 0$$

which implies

$$J(R\langle x\rangle) \subseteq (JR)\langle x\rangle. \quad \square$$

We recall that a ring R is called perfect if $R/J(R)$ is Artinian and $J(R)$ is T-nilpotent. Here, an ideal I is said to be T-nilpotent if given a sequence $\alpha = \{x_n\}$ of elements of I there exists a $j = j(\alpha)$ such that $x_1x_2\cdots x_j = 0$. In particular, choosing α a constant sequence we see that a T-nilpotent ideal is nil. A ring R is said to be semi-perfect if idempotents mod J(R) can be lifted and R/J(R) is Artinian. For example, an Artinian ring is perfect.

Corollary 2.3 (Parmenter-Sehgal [2]). If R is a perfect ring then $J(R\langle x\rangle) = (JR)\langle x\rangle$.

Proof: We have already seen that $J(R\langle x\rangle) \subseteq (JR)\langle x\rangle$. It is enough to check that $(JR)\langle x\rangle$ is nil. For this it suffices to observe that T-nilpotent implies locally nilpotent (Bass [1]). This can be seen as follows. Suppose that J is a T-nilpotent ideal and S is a finitely generated ring, $S = \langle x_1,\cdots,x_t\rangle$, $x_i \in J$. Suppose that S is not nilpotent. Therefore for every n we have a product $s_n = x_{i_1}x_{i_2}\cdots x_{i_n}$ of n x's which is not zero. We thus have a sequence $s_1,s_2,\cdots,s_n,\cdots$ of these nonzero products. Suppose that we have already picked a subsequence of $\{s_n\}$ all terms of which begin with $x_{\ell_1}x_{\ell_2}\cdots x_{\ell_{j-1}}$. Since we have only t choices for the next factor there exists a subsequence of the last sequence beginning with say $x_{\ell_1}x_{\ell_2}\cdots x_{\ell_j}$. Consequently we have found a sequence $x_{\ell_1},x_{\ell_2},\cdots,x_{\ell_j},\cdots$ such that no $x_{\ell_1}x_{\ell_2}\cdots x_{\ell_r} = 0$. Thus J is not T-nilpotent which is a contradiction proving the corollary. □

Corollary 2.4. R Artinian $\Rightarrow J(R\langle x\rangle) = (JR)\langle x\rangle$.

3. PERFECT RINGS

We shall prove in this section that for certain classes of rings the problem of this chapter can indeed be answered affirmatively. The next result is proved in (III.1.5).

Lemma 3.1. $(R_1 \oplus R_2)\langle x\rangle \simeq R_1\langle x\rangle \oplus R_2\langle x\rangle$.

Lemma 3.2. Let F and K be fields such that $\sigma\colon F\langle x\rangle \to K\langle x\rangle$ is an isomorphism. Then $\sigma(F) = K$.

Proof: The lemma is trivial if $|F| = |K| = 2$. Let $f \neq 0,-1$ be an element of F. Since $\sigma(f)$ is a unit of $K\langle x\rangle$, we have $\sigma(f) = kx^i$, $k \in K$. Therefore, $\sigma(1+f) = 1 + kx^i$. Since $1 + f$ is a unit, $i = 0$ and $\sigma(F) \subseteq K$. By using σ^{-1} we conclude that $\sigma(F) = K$. □

Lemma 3.3. Let R and S be finite direct sums of fields such that $\sigma\colon R\langle x\rangle \to S\langle x\rangle$ is an isomorphism. Then $\sigma(R) = S$.

Proof: Let $R = F_1 \oplus \cdots \oplus F_n$, $S = K_1 \oplus \cdots \oplus K_m$ be direct sums of fields F_i and K_j. With the identification of Lemma 3.1,

$$R\langle x\rangle = F_1\langle x\rangle \oplus \cdots \oplus F_n\langle x\rangle$$

and

$$S\langle x\rangle = K_1\langle x\rangle \oplus \cdots \oplus K_m\langle x\rangle.$$

The only primitive idempotents in $R\langle x\rangle$ are $e_i = (0,\cdots,0,1,0,\cdots,0)$ and similarly for $S\langle x\rangle$. Thus $n = m$ and

$$\sigma F_i\langle x\rangle = \sigma(e_i R\langle x\rangle) = (\sigma e_i)S\langle x\rangle = K_{j_i}\langle x\rangle$$

for some j_i. By Lemma 3.2 it follows that $\sigma(F_i) = K_{j_i}$ and hence $\sigma(R) = S$. □

Theorem 3.4 (Parmenter-Sehgal [2]). Suppose that for $i = 1,2$ R_i are rings with 1 such that

(i) Z_i the center of R_i is semi-perfect; and

(ii) $J(Z_i)$ is nil.

Then $R_1\langle x\rangle \simeq R_2\langle x\rangle \Rightarrow R_1 \simeq R_2$.

Proof: Let $\sigma\colon R_1\langle x\rangle \to R_2\langle x\rangle$ be the isomorphism. Then by restriction we have $\sigma\colon Z_1\langle x\rangle \to Z_2\langle x\rangle$. Due to the second condition of the hypothesis and (2.2) we have $J(Z_i\langle x\rangle) = (JZ_i)\langle x\rangle$. Thus there is the induced isomorphism

$$\bar{\sigma}\colon (Z_1/JZ_1)\langle x\rangle \to (Z_2/JZ_2)\langle x\rangle.$$

By Lemma 3.3 it follows that $\bar{\sigma}(Z_1/JZ_1) = (Z_2/JZ_2)$. Also, Lemma 3.1 gives

$$(Z_2/JZ_2)\langle x\rangle = \bar{e}_1(Z_2/JZ_2)\langle x\rangle \oplus \cdots \oplus \bar{e}_n(Z_2/JZ_2)\langle x\rangle,$$

where each $\bar{e}_i$ is a primitive idempotent and each $\bar{e}_i(Z_2/JZ_2)$ is a field F_i. Then we have

$$\bar{\sigma}(x) = \left(f_1 x^{i_1}, \cdots, f_n x^{i_n}\right) \quad \text{where } i_j = \pm 1,\quad 0 \neq f_j \in F_j$$

for all j. This follows because $\bar{\sigma}(Z_1/JZ_1)$ and $\bar{\sigma}(x)$ together must generate $(Z_2/JZ_2)\langle x\rangle$.

Since Z_2 is semi-perfect we can lift the idempotents $\bar{e}_i$ of Z_2/JZ_2 to primitive orthogonal idempotents e_i in Z_2 such that $1 = e_1 + \cdots + e_n$ (see Lambek [1]). Then

$$R_2\langle x\rangle = e_1R_2\langle x\rangle \oplus \cdots \oplus e_nR_2\langle x\rangle.$$

Define an R_2-automorphism of $R_2\langle x\rangle$ by $\beta(x) = \left(x^{i_1}, x^{i_2}, \cdots, x^{i_n}\right)$. This induces an automorphism $\bar{\beta}$ of $(Z_2/JZ_2)\langle x\rangle$. Notice that

$$\bar{\beta}\,\bar{\sigma}(x) = (f_1x, f_2x, \cdots, f_nx) = (f_1, \cdots, f_n)x = ux$$

where u is a unit of Z_2/JZ_2. Hence, we see that

$$\beta\,\sigma(x) = u_1x + \sum_{i\neq 1} a_ix^i \qquad (*)$$

where u_1 is a unit of Z_2 and a_i are nilpotent elements of Z_2. By Lemma 1.4 we have an automorphism α of $Z_2\langle x\rangle$ given by $\alpha(x) = \beta\sigma(x)$. Extend α R_2-linearly to $R_2\langle x\rangle$. Clearly α is onto. To see that α is 1-1, suppose that

$$\sum_{j>0} b_j \alpha(x)^j = 0, \quad b_j \in R_2. \tag{**}$$

Let A be the ideal of R_2 generated by the central nilpotent elements, a_i, $i \neq 1$. Then A is nilpotent. We have

$$\sum b_j u_1^j x^j \equiv 0 \pmod{A\langle x\rangle}$$

and thus $b_j \in A$. Suppose that all $b_j \in A^k$, $k \geq 1$ for all j. Then from (*) and (**) we get

$$\sum b_j u_1^j x^j \equiv 0 \pmod{A^{k+1}\langle x\rangle}$$

and consequently $b_j \in A^{k+1}$. It follows from the nilpotence of A that $b_j = 0$ for all j. We have proved that $x \to \beta\sigma(x)$ induces an R_2-automorphism α of $R_2\langle x\rangle$. This gives a ring isomorphism

$$\alpha^{-1}\beta\sigma\colon R_1\langle x\rangle \to R_2\langle x\rangle$$

such that $(\alpha^{-1}\beta\sigma)(x) = x$ and therefore $\alpha^{-1}\beta\sigma\colon \Delta(R_1\langle x\rangle) \to \Delta(R_2\langle x\rangle)$. Hence,

$$R_1 \simeq R_1\langle x\rangle/\Delta(R_1\langle x\rangle) \simeq R_2\langle x\rangle/\Delta(R_2\langle x\rangle) \simeq R_2. \quad \square$$

Corollary 3.5. Let R_1, R_2 be rings with perfect centers; then

$$R_1\langle x\rangle \simeq R_2\langle x\rangle \Rightarrow R_1 \simeq R_2.$$

Corollary 3.6. Let R_1, R_2 be rings with Artinian centers; then

$$R_1\langle x\rangle \simeq R_2\langle x\rangle \Rightarrow R_1 \simeq R_2.$$

We like to remark that there exists an Artinian ring whose center is not Artinian and there exists a non-Artinian ring whose center is Artinian.

Example 1. Let $R = F[x,y]$ be the ring of polynomials in two non-commuting variables x, y over a field F. Then R is not Artinian but its center, F, is Artinian.

Example 2 (Björk). Let K be a field with a derivation $f: K \to K$ such that if $K_o = \{k \in K: f(k) = 0\}$ then the dimension of K/K_o is infinite. Let $R = \{Ke + Kx + Ky$: e is the unit of R, $x^2 = xy = yx = y^2 = 0$, $ky = yk$, $xk = kx + f(k)y$ for all $k \in K\}$. Then the center of R is $K_oe + Ky$ which is easily seen to be non-Artinian while R is Artinian.

Essentially the same arguments as in Theorem 3.4 give the next theorem.

Theorem 3.7. Let R_1 and R_2 be perfect rings. Then

$$R_1\langle x\rangle \overset{\sigma}{\simeq} R_2\langle x\rangle \Rightarrow R_1 \simeq R_2.$$

Proof: (Sketch) Let us denote by $Z(R)$ the center of the ring R. Since by (2.3) $J(R_1\langle x\rangle) = (JR_1)\langle x\rangle$ we have the induced isomorphism

$$\bar{\sigma}: Z\big((R_1/JR_1)\big)\langle x\rangle \to Z\big((R_2/JR_2)\big)\langle x\rangle.$$

Writing $Z(R_2/JR_2)$ as a direct sum of fields

$$Z(R_2/JR_2) = F_1 \oplus \cdots \oplus F_n , \tag{*}$$

we have

$$\bar{\sigma}(x) = \big(f_1x^{i_1},\cdots,f_nx^{i_n}\big), \quad o \neq f_j \in F_j, \quad i_j = \pm 1.$$

Lifting the idempotents $\bar{e}_i$, in the decomposition (*), to orthogonal idempotents e_i of R_2 we get

$$R_2\langle x\rangle = e_1R_2\langle x\rangle \oplus \cdots \oplus e_nR_2\langle x\rangle .$$

Define an R_2-automorphism β of $R_2\langle x\rangle$ by

$$\beta(x) = \big(x^{i_1},\cdots,x^{i_n}\big).$$

This induces an automorphism $\bar{\beta}$ of $(R_2/JR_2)\langle x\rangle$ and

$$\overline{\beta}\,\overline{\sigma}(x) = (f_1x,\cdots,f_nx) = ux$$

where u is a unit of R_2/JR_2. We conclude that

$$\beta\,\sigma(x) = u_1x + \sum_{i\neq 1} a_ix^i$$

where u_1 is a unit, $a_i \in JR_2$ which is nil. But since x is central we conclude that u_1 and $a_i \in Z(R_2)$. Now let α be the R_2-automorphism of $R_2\langle x\rangle$ given by

$$\alpha\colon x \to \beta\sigma(x).$$

Thus we have an isomorphism

$$\alpha^{-1}\beta\sigma\colon R_1\langle x\rangle \to R_2\langle x\rangle$$

such that $(\alpha^{-1}\beta\sigma)(x) = x$ and consequently $R_1 \simeq R_2$. □

Corollary 3.8. Let R_1, R_2 be Artinian rings; then

$$R_1\langle x\rangle \simeq R_2\langle x\rangle \Rightarrow R_1 \simeq R_2.$$

4. COMMUTATIVE RINGS

Recall that a ring R with 1 is called local if the non-units of R form an ideal. R is said to be (von Neumann) regular if for any $a \in R$, there exists $b \in R$ with $aba = a$.

Theorem 4.1 (Parmenter [3]). Let R_1 and R_2 be commutative regular; then

$$R_1\langle x\rangle \overset{\sigma}{\simeq} R_2\langle x\rangle \Rightarrow R_1 \overset{\sigma}{\simeq} R_2.$$

Proof: Given $f \in R_1\langle x\rangle$. Suppose there exist $g,h \in R_1\langle x\rangle$ such that $fgf = f$ and $(f-1)h(f-1) = (f-1)$. Let P be a prime ideal of R_1. We claim that there is a $c \in R_1$ such that $f-c \in P\langle x\rangle$.

In $(R_1/P)\langle x\rangle$ we have

$$(\overline{f}\,\overline{g}-\overline{1})\overline{f} = 0 \quad \text{and} \quad \left((\overline{f-1})\overline{h}-\overline{1}\right)(\overline{f-1}) = 0.$$

Either one of $\overline{f}$ or $\overline{f-1} = 0$ in which case the claim is established or we deduce that $\overline{f}\,\overline{g} = 1$, $(\overline{f-1})\overline{h} = 1$. Since the units of $R_1/P\langle x\rangle$ are trivial it follows that $\overline{f} = \overline{c}$, $c \in R_1$. In any case we can write

$$f = c_0 + \sum_{i \neq 0} c_i x^i \qquad c_i \in P\text{, for all } i \neq 0.$$

Since $c_i \in P$ for all prime ideals P it follows that c_i are nilpotent for $i \neq 0$. But R_1, being regular, has no nonzero nilpotent elements. Thus $f \in R_1$. We have, therefore, characterized the ring R_1 to be consisting of those elements $f \in R_1\langle x\rangle$ for which there exist elements g and $h \in R_1\langle x\rangle$ satisfying $fgf = f$, $(f-1)h(f-1) = f-1$. Hence $\sigma(R_1) = R_2$. □

<u>Theorem 4.2</u> (Parmenter [3]). Let R_1 and R_2 be commutative local rings; then

$$R_1\langle x\rangle \overset{\sigma}{\simeq} R_2\langle x\rangle \Rightarrow R_1 \simeq R_2.$$

<u>Proof</u>: Suppose that R_1 (and therefore R_2) has no nonzero nilpotent elements. Being local R_1 and R_2 certainly have no idempotents. Thus by (1.0) we have

$$U(R_i\langle x\rangle) = (UR_i)\cdot\langle x\rangle, \qquad i = 1,2.$$

Let r be a unit of R_1; then $\sigma(r) = ux^i$, $u \in UR_2$, $i \in \mathbb{Z}$ and $\sigma(r^{-1}) = u^{-1}x^{-i}$. As a result $\sigma(r+r^{-1}) = ux^i + u^{-1}x^{-i}$ and $\sigma(r+r^{-1}-1) = ux^i + u^{-1}x^{-i}-1$. Since one of $r + r^{-1}$ and $(r+r^{-1}-1)$ is a unit of R_1 it follows that $i = 0$ and $\sigma(R_1) \subseteq R_2$. We have proved:

(4.3) $\quad R_1$ has no nonzero nilpotent elements $\Rightarrow \sigma(R_1) \subseteq R_2$.

Denote by N_i the ideal of nilpotent elements of R_i. Then R_i/N_i is also local. Consequently, by (1.0) we have

$$\mathcal{U}(R_i/N_i)\langle x\rangle = \mathcal{U}(R_i/N_i)\cdot\langle x\rangle.$$

Also, (4.3) implies that $\sigma(R_1/N_1) \subseteq R_2/N_2$. It follows that

$$(4.4)\qquad \sigma(x) = ux^{\varepsilon} + \sum_{j\neq\varepsilon} n_j x^j, \quad \varepsilon = \pm 1, \quad n_j \in N_2, \quad u \in \mathcal{U}R_2.$$

Lemma 1.4 implies that $R_2\langle x\rangle = R_2\langle\sigma(x)\rangle$. Therefore, $R_1\langle x\rangle \simeq R_2\langle x\rangle = R_2\langle\sigma(x)\rangle$. Thus,

$$\begin{aligned}\sigma(\Delta_{R_1}\langle x\rangle) &= \sigma\big((1-x)R_1\langle x\rangle\big) = \big(1-\sigma(x)\big)R_2\langle x\rangle = \big(1-\sigma(x)\big)R_2\langle(\sigma x)\rangle\\ &= \Delta_{R_2}\langle\sigma(x)\rangle\end{aligned}$$

and

$$R_1 \simeq R_1\langle x\rangle/\Delta_{R_1}\langle x\rangle \simeq R_2\langle\sigma(x)\rangle/\Delta_{R_2}\langle\sigma(x)\rangle \simeq R_2. \quad \square$$

<u>Theorem 4.5</u> (Parmenter). Let R_1 and R_2 be commutative. Then $R_1\langle x\rangle \simeq R_2\langle x\rangle \Rightarrow R_1$ and R_2 are subisomorphic.

We need the following lemma.

<u>Lemma 4.6</u>. Let R be commutative and let $u = \Sigma a_i x^i \in R\langle x\rangle$ be a unit with a_0 nilpotent. Then $R \cap R\langle x\rangle(u-1) = 0$.

<u>Proof</u>: Write $u = \Sigma a_i x^i$, $u^{-1} = \Sigma b_i x^i$. Then by (1.2) we have

$$\sum a_i b_{-i} = 1, \quad a_i a_j, b_i b_j, a_r b_s \text{ nilpotent for } i \neq j, \ r \neq -s. \qquad (*)$$

Let $R \ni r = g(x)(\Sigma a_i x^i - 1)$, $g(x) = \Sigma g_i x^i$, $g_i \in R$. For any prime ideal P of R we get in $R/P\langle x\rangle$

$$\bar{r} = \left(\sum \bar{g}_i x^i\right)\left(\bar{a}_{i_0} x^{i_0} - 1\right)$$

for some i_0 as, R/P being an integral domain, units of $R/P\langle x\rangle$ are trivial.

Comparing the coefficients of the highest degree terms on both sides we conclude that $\bar{g}_i = 0$ for all i, i.e. $g_i \in P$. It follows that all g_i are nilpotent. Let I be the ideal generated by $\{g_i\} \cup \{a_i : a_i \text{ nilpotent}\} \cup \{a_i a_j : i \neq j\}$. Then I is nilpotent. Suppose that all $g_i \in I^k$ for $k \geq 1$. Consider the equation

$$r = \left(\sum g_i x^i\right)\left(\sum a_i x^i - 1\right).$$

Suppose that a_n is not nilpotent. Then

$$a_n r = \left(\sum g_i x^i\right)(a_n^2 x^n - a_n) \pmod{I^{k+1}\langle x\rangle}.$$

If s is the largest i such that $g_i \notin I^{k+1}$ we get $a_n^2 g_s \equiv 0$ or $a_n g_s \equiv 0 \pmod{I^{k+1}}$. But (*) implies that we have

$$a_n g_s \equiv a_n g_s \left(\sum a_i b_{-i}\right) \equiv a_n^2 b_{-n} g_s \equiv 0 \pmod{I^{k+1}}.$$

Again, multiplying the equation $\Sigma a_i b_{-i} = 1$ by g_s we conclude that $g_s \in I^{k+1}$. Thus $g_i \in I^{k+1}$ for all i and it follows that $r = 0$.

4.7. <u>Proof of Theorem 4.5</u>

Let σ be the given isomorphism $R_1\langle x\rangle \to R_2\langle x\rangle$. R_1 is isomorphic to $R_1^\sigma = S$. Let $u = \sigma(x) = \Sigma a_i x^i$. Then $R_2\langle x\rangle = S\langle u\rangle$. So it suffices to prove that, if $R\langle x\rangle = S\langle u\rangle$ then R and S are subisomorphic.

Suppose that a_0 is nilpotent (perhaps zero). Then, by (1.3), $\{u^j : j \in \mathbb{Z}\}$ is linearly independent over R. So we can regard $R\langle u\rangle$ as a subring of $R\langle x\rangle$. Let ρ be the augmentation map $S\langle u\rangle \to S$. Then

$$\operatorname{Ker} \rho = \Delta_S\langle u\rangle = S\langle u\rangle(u-1) = R\langle x\rangle(u-1).$$

Clearly,

$$\Delta_R\langle u\rangle = R\langle u\rangle(1-u) \subseteq \operatorname{Ker} \rho.$$

Also, we know by the last lemma that $R \cap \operatorname{Ker} \rho = 0$. It follows that

$$R\langle u\rangle \cap \operatorname{Ker} \rho = \Delta_R\langle u\rangle.$$

We have, therefore,

$$R \simeq R\langle u\rangle/\Delta_R\langle u\rangle$$

is imbedded in S.

It remains to consider the case when a_0 is not nilpotent, i.e. $R\langle x\rangle = S\langle u\rangle$, $u = \Sigma a_i x^i$, a_0 not nilpotent. Let us write

$$u^{-1} = \sum b_j x^j, \quad b_j \in R.$$

Then by (1.2) we get

$$\sum a_i b_{-i} = 1, \quad a_i b_j \text{ nilpotent for } i \neq -j.$$

Thus $a_0 b_0 a_0 - a_0$ is nilpotent and $(a_0 b_0)^2 = a_0 b_0 + n$ where n is nilpotent. Hence there exists an idempotent $e \in R$ such that $e - a_0 b_0$ is nilpotent (see Lambek [1], page 72). It follows by (I.4.4) that $e \in S$. Then

$$e \cdot R\langle x\rangle = (eR)\langle x\rangle = (eR)\langle ex\rangle$$

$$e \cdot S\left\langle \sum a_i x^i\right\rangle = (eS)\left\langle \sum a_i x^i\right\rangle = (eS)\left\langle e\sum a_i x^i\right\rangle .$$

Similarly for the idempotent $1 - e$. But the constant term of $(1-e)\Sigma a_i x^i$ is $(1-e)a_0 = a_0 - ea_0$. Since $e - a_0 b_0$ is nilpotent it follows that $ea_0 - a_0 b_0 a_0$ is nilpotent. But also as remarked earlier $a_0 b_0 a_0 - a_0$ is nilpotent. Thus $(1-e)a_0$ is nilpotent. Hence by the case already considered $(1-e)R$ is imbedded in $(1-e)S$. It remains to conclude from the assumption $eR\langle x\rangle = eS\langle \Sigma a_i x^i\rangle$, ea_0 not nilpotent, that eR can be imbedded in eS. Notice that since $e - a_0 b_0$ is nilpotent, ea_i is nilpotent for each $i \neq 0$ and since $(1-e)a_0$ is nilpotent, ea_0 is a unit of eR. Therefore it is enough to prove:

(4.8) If $R\langle x\rangle = T\left\langle \sum a_i x^i\right\rangle$, a_0 a unit, a_i nilpotent $i \neq 0$ then R can be imbedded in T.

Let us write $u = \Sigma a_i x^i$. Then we have $R\langle x\rangle = T\langle u\rangle$ and x being a unit of T<u> we have an expression

$$x = \sum_j c_j u^j, \quad c_j \in T.$$

Moreover, (1.2) implies that there exist elements $d_i \in T$ such that

$$\sum_j c_j d_{-j} = 1, \quad c_j d_k, c_r c_\ell, d_r d_\ell \text{ nilpotent for } j \neq -k, \; r \neq \ell. \qquad (*)$$

Consider, for a fixed non-nilpotent c_k, the element

$$v = c_k u + \left(\sum_{\ell \neq k} c_\ell\right) u^{-1}$$

of T<u>. This element is a unit because, if we denote by N the ideal of nilpotent elements of T, we have

$$\left(c_k u + \left(\sum_{\ell \neq k} c_\ell\right) u^{-1}\right)\left(\left(\sum_{\ell \neq -k} d_\ell\right) u + d_{-k} u^{-1}\right) \equiv c_k d_{-k} + \left(\sum_{\ell \neq k} c_\ell\right)\left(\sum_{\ell \neq -k} d_\ell\right)$$

$$\equiv \sum_i c_i d_{-i} \equiv 1 \pmod{N\langle u\rangle}.$$

Further, we wish to prove that if we write $v = \Sigma r_i x^i$, $r_i \in R$ then r_0 is nilpotent. Suppose that P is a prime ideal of R such that a certain $c_s \notin P\langle x\rangle$. For $t \neq s$, $c_s c_t$ is nilpotent by (*) and thus belongs to the prime ideal P<x>. It follows that $c_t \in P\langle x\rangle$. Therefore, since $u \equiv a_0 \bmod P$,

$$x \equiv c_s u^s \equiv c_s a_0^s \pmod{P\langle x\rangle}.$$

Thus

$$c_s = x a_0^{-s} + \sum m_{i,s} x^i, \quad m_{i,s} \in P \subset R. \qquad (**)$$

Such an expression exists for c_s whenever there is a prime ideal P of R such that $c_s \notin P\langle x\rangle$. It follows that $m_{i,s}$ for $i \neq 1$ is nilpotent because otherwise there exists a prime ideal Q of R such that $m_{i,s} \notin Q$ and consequently $c_s \notin Q\langle x\rangle$ which contradicts (**). We have proved the following:

If c_s is not nilpotent then $c_s = xd_s + \Sigma m_{i,s} x^i$ where $m_{i,s}$ are nilpotent elements of R and $d_s \in R$.

Consider v modulo $I\langle x\rangle$ where I is a prime ideal of R. All the nilpotent $c_i \in I\langle x\rangle$ as $I\langle x\rangle$ is a prime ideal of $R\langle x\rangle$. We have

$$\begin{aligned} \sum r_i x^i = v &= c_k u + \left(\sum_{\ell \neq k} c_\ell\right) u^{-1} \\ &\equiv c_k a_0 + \left(\sum_{\text{some } \ell} c_\ell\right)(a_0^{-1}) \pmod{I\langle x\rangle} \\ &\equiv x d_k a_0 + \delta x \bmod I\langle x\rangle, \quad \delta \in R \quad (\text{by } (**)). \end{aligned}$$

It follows that $r_0 \in I$. Since I is an arbitrary prime ideal of R we deduce that r_0 is nilpotent. Thus we have by (1.4)

$$R\langle x\rangle = T\langle u\rangle = T\langle v\rangle$$

where $v = \Sigma r_i x^i$ with r_0 nilpotent. We have reduced (4.8) to the case already settled.

To sum up, we have proved that if $R\langle x\rangle = S\langle u\rangle$ then R can be imbedded in S. By symmetry S can also be imbedded in R. □

CHAPTER V

LIE PROPERTIES IN KG

1. PRELIMINARIES

Any associative ring R may be regarded as a Lie ring by defining the Lie multiplication by $[a,b] = ab - ba$, for $a,b \in R$. Let R be a k-algebra (with or without a 1) over a commutative ring k with identity. For any two k-subspaces A and B of R, we define $[A,B]$ to be the additive subgroup of R generated by all the Lie products $[a,b] = ab - ba$ with $a \in A$, $b \in B$. Clearly, $[A,B]$ is a k-subspace of R. We can then define inductively the Lie central and Lie derived series of R by

$$R^{[1]} = R, \quad R^{[n+1]} = [R^{[n]},R]$$

and

$$\delta^{[0]}R = R, \quad \delta^{[n+1]}R = [\delta^{[n]}R,\delta^{[n]}R]$$

respectively. One may also enlarge the terms of this series by making them associative ideals at each stage. More precisely, we define by induction the series

$R^{(1)} = R$, $R^{(n+1)}$ = the (associative) ideal of R generated by $[R^{(n)},R]$

and

$\delta^{(0)}R = R$, $\delta^{(n+1)}R$ = the (associative) ideal of R generated by $[\delta^{(n)}R,\delta^{(n)}R]$.

These series have been investigated by Jennings ([2],[4]) and Sandling ([1],[2]).

We say that R is Lie nilpotent if $R^{[n]} = 0$ for some integer n and similarly R is Lie solvable if $\delta^{[n]}R = 0$ for some integer n. We also say that R is strongly Lie nilpotent (resp. strongly Lie solvable) if $R^{(n)} = 0$ (resp. $\delta^{(n)}R = 0$) for some integer n. Let us recall that R is said to be Lie n-Engel if it satisfies for all $x,y \in R$ the identity

$$[x,\underbrace{y,y,\cdots,y}_{n}] = 0.$$

Here, by $[x_1,x_2,\cdots,x_n]$ we understand $\Big[[x_1,\cdots,x_{n-1}],x_n\Big]$.

In this chapter we intend to investigate as to when the group ring kG is Lie solvable, Lie nilpotent or Lie n-Engel. One approach to those questions which has moderate success is via a well known theorem of Herstein ([1], page 11). We shall however use a theorem of Passman on group rings satisfying polynomial identities. Let $K[z_1,\cdots,z_n]$ be the polynomial ring over the field K in the non-commuting variables $z_1,\cdots,z_n$. An algebra R over K is said to satisfy a polynomial identity if there exists $f(z_1,\cdots,z_n) \in K[z_1,\cdots,z_n]$, $f \neq 0$ with

$$f(\alpha_1,\cdots,\alpha_n) = 0$$

for all $\alpha_1,\cdots,\alpha_n \in R$. It is apparent that R is Lie nilpotent or Lie solvable if and only if it satisfies certain (multilinear) polynomial identities. Thus, for example, $\delta^{[2]}R = 0$ if and only if R satisfies the identity

$$\Big[[z_1,z_2],[z_3,z_4]\Big] .$$

In order to state Passman's theorem we need to recall the definition of the FC-subgroup $\phi(G)$ of a group G, namely,

$$\phi(G) = \{g \in G\colon\ g \text{ has a finite number of conjugates in } G\}.$$

Theorem 1.1 (Passman [4]). Let KG, the group algebra of G over a field K, satisfy a polynomial identity. Then $(G: \phi(G)) < \infty$ and the commutator group $\phi(G)'$ is finite.

2. MATRIX RINGS

Let S be a commutative K-algebra (with or without a 1), where K is a field. Suppose that the characteristic of K, char K, is $p \geq 0$. Then we let $M_n(S)$ denote the ring of $n \times n$ matrices over S and let $T_n(S)$ be those matrices which have trace zero. We shall denote by $\{e_{ij}\}$ the set of matrix units, namely, e_{ij} is the matrix with 1 in the (i,j) place and zeros elsewhere.

Lemma 2.1. With S as above we have

(i) $\quad [T_n(S),M_n(S)] = T_n(S^2)$;

(ii) $\quad [T_n(S),T_n(S)] = T_n(S^2)$ unless $n = 2$ and $p = 2$.

Proof: If $n = 1$ then $T_n(S) = \{0\}$ so that the result is clear. Thus let $n \geq 2$. Clearly,

$$[M_n(S),M_n(S)] \subseteq T_n(S^2),$$

so we need only prove the reverse inclusions.

Let $i \neq j$. If $n \geq 3$, we can choose $\ell \neq i,j$. Then for any $s,s' \in S$ we have

$$[se_{ij},s'(e_{jj}-e_{\ell\ell})] = ss'e_{ij}$$

$$[se_{ij},s'e_{ji}] = ss'(e_{ii}-e_{jj}).$$

Since all these matrices have trace 0 and since the right hand matrices span $T_n(S^2)$, part (ii) and therefore part (i) follows.

Now let $n = 2$. From the above and

$$[se_{ij},s'e_{jj}] = ss'e_{ij}$$

we see that (i) holds. Finally if $p \neq 2$ then

$$[se_{ij}, s'(e_{jj}-e_{ii})] = 2ss'e_{ij}$$

implies that $ss'e_{ij} \in [T_n(S), T_n(S)]$ and the result follows. □

Lemma 2.2. Assume that S is not nilpotent. Then

(i) $M_n(S)$ is Lie nilpotent if and only if $n = 1$;

(ii) $M_n(S)$ is Lie solvable if and only if $n = 1$ or $n = 2$, $p = 2$.

Proof: If $n = 1$, then $M_n(S)$ is commutative and hence both Lie nilpotent and Lie solvable. Let $n > 1$. Then by Lemma 2.1(i) and induction we have for $\ell \geq 1$

$$\left(M_n(S)\right)^{[\ell]} \supseteq T_n(S^{2^\ell}).$$

Since S is not nilpotent, $T_n(S^{2^\ell}) \neq 0$ and $M_n(S)$ is not Lie nilpotent. Moreover, by Lemma 2.1(ii) unless $n = 2$ and $p = 2$,we have for all $\ell \geq 2$

$$\delta^{[\ell]}\left(M_n(S)\right) \supseteq T_n(S^{2^\ell}) \neq 0.$$

Thus $M_n(S)$ is not Lie solvable.

Finally let $n = 2$ and char $K = 2$. Then each matrix in $\delta^{[1]}M_2(S)$ lies in the span of

$$S^2e_{12},\quad S^2e_{21} \quad \text{and} \quad S^2(e_{11}+e_{12}).$$

Thus since $e_{11} + e_{22}$, the identity matrix, is central $\delta^{[2]}M_2(S)$ is contained in

$$[S^2e_{12}, S^2e_{21}] = S^4(e_{11}+e_{22})$$

and hence $\delta^{[3]}M_2(S) = 0$. Therefore $M_2(S)$ is Lie solvable if $p = 2$. □

Let A be a normal Abelian subgroup of G of finite index n.

Choose, for convenience, $x_1 = 1, x_2^{-1}, \cdots, x_n^{-1}$ a set of coset representatives. $KG = \Sigma_i KAx_i^{-1}$ is free as a left KA-module on the basis of coset representatives. An element α of KG induces a (nontrivial) homomorphism from KG to itself by right multiplication. Since right and left multiplication commute as operators on KG, this is a KA-homomorphism and so is represented by an $n \times n$ KA-matrix according to the given basis; explicitly, we shall use the notation α_{ij} for the entries of the matrix corresponding to α so that

$$x_i^{-1}\alpha = \sum_j \alpha_{ij} x_j^{-1}, \qquad \alpha_{ij} \in KA.$$

We have thus defined a homomorphism

$$\rho\colon KG \to M_n(KA).$$

By the nontriviality, we see that

(2.3) ρ is a monomorphism.

For each $i = 2,3,\cdots,n$ the commutators

$$(A,x_i) = \{a^{-1}x_i^{-1}ax_i : a \in A\}$$

form a subgroup isomorphic to $A/C_A(x_i)$ where $C_A(x_i)$ is the centralizer of x_i in A. This can be seen by considering the map

$$A \ni a \to (a,x_i) \in (A,x_i)$$

which is a homomorphism due to the commutativity of A.

Let $S_i = \Delta(A,x_i)$ be the augmentation ideal of the group ring $K(A,x_i)$. Observe that KA is embedded in $M_n(KA)$ in two different ways. First, $KA \subseteq M_n(KA)$ consists of all scalar matrices. Secondly, $\rho(KA) \subseteq M_n(KA)$. These are easily seen to be diagonal matrices but not necessarily scalars.

<u>Lemma 2.4</u>. Let S_i be as above. Set $S = S_2S_3\cdots S_n$.

(i) If $x_i \notin \phi(G)$, for $i = 2,\cdots,n$ then S is not nilpotent.

(ii) Let R be the subring $R = (KA)\rho(KG)$ of $M_n(KA)$. Then $R \supseteq M_n(S)$.

Proof: (i) Since $x_i \notin \phi(G)$, $(G: C_G(x_i))$ is infinite and thus $(A: C_A(x_i))$ is infinite. It follows that (A,x_i) is an infinite group. Thus by (III.4.18), S_i annihilates no nonzero element of KA. As a consequence S is not nilpotent.

(ii) Let us make a couple of observations first. Suppose $\alpha \in KA$. Then $x_i^{-1}\alpha = \alpha^{x_i}x_i^{-1}$. It follows that $\rho(\alpha)$ is a diagonal matrix. Namely,

$$\text{(2.5)} \qquad \alpha \in KA \Rightarrow \rho(\alpha) = \operatorname{diag}\left(\alpha^{x_1},\alpha^{x_2},\cdots,\alpha^{x_n}\right) .$$

We claim that

$$\text{(2.6)} \qquad e_{11}\rho(x_i^{-1}) = e_{1i}, \quad \rho(x_i)e_{11} = e_{i1}.$$

Indeed, $x_1^{-1}x_i^{-1} = x_i^{-1}$ implies that the first row of the matrix $\rho(x_i^{-1})$ is e_{1i} and $e_{11}\rho(x_i^{-1}) = e_{11}e_{1i} = e_{1i}$. Similarly, the first column of the matrix $\rho(x_i)$ is precisely e_{i1} and $\rho(x_i)e_{11} = e_{i1}e_1 = e_{i1}$.

Now let $a \in A$ and let $i \geq 2$. Then $a^{-1}\left(a^{x_i} - \rho(a)\right) \in R$. It follows by (2.5) that this is a diagonal matrix whose j-th diagonal entry is $a^{-1}\left(a^{x_i}-a^{x_j}\right)$. In particular this is zero for $j = i$ and for $j = 1$ we have

$$a^{-1}\left(a^{x_i}-a\right) = (a,x_i) - 1.$$

Thus we conclude that for any element $g \in (A,x_i)$, R contains a matrix of the form

$$\operatorname{diag}(g-1,*,0,*),$$

where 0 is in the i-th position. Thus for any element $s_i \in S_i$, R contains a matrix of the form

$$\alpha_i = \operatorname{diag}(s_i,*,0,*).$$

Then $\alpha = \alpha_2\alpha_3\cdots\alpha_n = \mathrm{diag}(s_2s_3\cdots s_n, 0,\cdots,0) \in R$. Thus R contains Se_{11}. Finally by (2.6) for any i,j

$$R \supseteq \rho(x_i)Se_{11}\rho(x_j^{-1}) = Se_{i1}e_{1j} = Se_{ij}$$

which completes the proof of (ii). □

3. LIE SOLVABILITY

In this section we characterize group algebras KG which are Lie solvable. If p is a prime we say that G is p-abelian if its derived group G' is a finite p-group. We also say that G is 0-abelian if and only if G is Abelian.

Theorem 3.1 (Passi-Passman-Sehgal [1]). Necessary and sufficient conditions for the Lie solvability of KG, the group algebra of G over the field K with $\mathrm{char}\, K = p \geq 0$, are:

(i) G is p-abelian when $p \neq 2$,
(ii) G has a 2-abelian subgroup of index at most 2 when $p = 2$.

Proof: For the necessity part of $p = 2$ case we refer to the original paper as it involves a lot of finite group theory far removed from the rest of this book. The sufficiency for $p = 2$ is proved in (3.3). We shall prove below the $p \neq 2$ case. Let us first assume that $p > 0$, $p \neq 2$ and that KG is Lie solvable. By Theorem 1.1 we have

$$\big(G: \phi(G)\big) = n < \infty \quad \text{and} \quad |\phi(G)'| < \infty.$$

We assert that

(3.2) $\quad n = 1$ i.e. G is an FC-group.

Writing $W = \phi(G)'$, $\overline{G} = G/W$ it follows that

$$\phi(\overline{G}) = \phi(G)/W, \quad \big(G: \phi(G)\big) = \big(\overline{G}: \phi(\overline{G})\big) .$$

Since $K\overline{G}$ is Lie solvable, in order to prove (3.2) we may replace G by $\overline{G}$ and assume that $\phi(G)$ is Abelian. Now we have

$$\phi(G) = A \quad \text{Abelian}, \quad (G\colon A) = n < \infty.$$

Then by (2.4), in the notation set forth in the last section,

$$(KA)\rho(KG) = R \supseteq M_n(S).$$

Also by (2.4)(i), S is not nilpotent. But R is Lie solvable as KA is central and hence $M_n(S)$ is also Lie solvable. It follows by Lemma 2.2 that $n = 1$.

We have proved that G is an FC-group with $|G'| < \infty$. In order to prove that G' is a p-group we can assume that G is finitely generated, therefore, residually finite (I.4.6). Thus we may further assume that G is finite. Let $1 \neq g \in G'$ and suppose that $g-1 \notin JKG$, the Jacobson radical of KG. Then there exists an irreducible representation ψ of KG such that $\psi(g-1) \neq 0$. But $\psi(KG)$ is a finite dimensional simple algebra over K and hence $\psi(KG) \simeq M_m(K)$ for some m because we may assume without loss of generality that K is algebraically closed. Since the Lie solvability of KG implies that $\psi(KG)$ is Lie solvable it follows by Lemma 2.2 that $m = 1$. Thus

$$\psi(KG) \simeq K, \quad \psi(g-1) \neq 0, \quad g \in G'.$$

This is a contradiction as K is commutative. We have proved that $\Delta(G') \subseteq JKG$ and thus $\Delta(G')$ is nilpotent. It follows by (I.2.21) that G' is a p-group. Thus G is p-abelian.

Now let us suppose that $\operatorname{char} K = 0$ and that KG is Lie solvable. Then $\mathbb{Z}G \subseteq KG$ is Lie solvable and so is $(\mathbb{Z}/p\mathbb{Z})G$ for every prime p. It follows that G is p-abelian for all odd primes p and thus G is Abelian. This completes the proof of the necessity part of (i).

For the proof of sufficiency, we observe that if $\operatorname{char} K = 0$ then G is Abelian which certainly implies Lie solvability. The $p > 0$ case follows from the next lemma.

Lemma 3.3. Suppose that k is a commutative ring with 1 of characteristic $\chi > 0$. Then

(i) If χ is a power of a prime p and G is p-abelian then kG is strongly Lie solvable;

(ii) If χ is a power of 2 and G has a 2-abelian subgroup of index at most 2 then kG is Lie solvable.

Proof: (i) Since $k(G/G')$ is Abelian, $\delta^{(1)}(kG)$ is contained in the kernel of the map $kG \to k(G/G')$, that is,

$$\delta^{(1)}(kG) \subseteq \Delta(G,G') = \Delta(G')kG.$$

Consequently,

$$\delta^{(i+1)}(kG) \subseteq \left(\Delta(G')\right)^{2^i} kG = 0$$

for a suitable i as $\Delta(G')$ is nilpotent by (I.2.21).

(ii) In view of (i) we may assume that G has a 2-abelian subgroup N of index 2 and χ is a 2-power. Since $\Delta(G,N')$ is nilpotent we may factor by N' and assume that N is Abelian. Then since (2.3) applies to kG we have $kG \subseteq M_2(kN)$. Since $(k/2k)N$ is a $\mathbb{Z}/2\mathbb{Z}$-algebra and since $(k/2k)G \subseteq M_2\left((k/2k)N\right)$ we can conclude by (2.2) that there exists an r such that $\delta^{[r]}(kG) \subseteq 2kG$. It follows that for a suitable m, $\delta^{[m]}(kG) = 0$. □

The argument of Theorem 3.1 also proves the next result.

Lemma 3.4. If the group algebra KG of G over a field K of characteristic $p \geq 0$ is Lie nilpotent then G is p-abelian.

4. LIE NILPOTENCE

Before characterizing Lie nilpotent group rings kG we prove two results that we need.

Lemma 4.1. Let A be a normal Abelian subgroup of G. Suppose that A has exponent p^n and that G acts by conjugation on A as a finite p-group of automorphisms. Then

$$(A,G,G,\cdots,G) = (A,{}_rG) = 1 \quad \text{for a suitable } r.$$

Proof: We regard A as a $(\mathbb{Z}/p^n\mathbb{Z})G$-module by setting for $a \in A$, $g,h \in G$:

$$a^g = g^{-1}ag, \qquad a^\alpha = \prod\left(a^{\alpha(g)}\right)^g, \quad \text{where} \quad \alpha = \sum\alpha(g)g.$$

We have

$$(a,g) = a^{-1}g^{-1}ag = a^{-1}a^g = a^{(g-1)}$$

$$(a,gh) = a^{(g-1)}a^{(h-1)} = a^{(g-1)+(h-1)}$$

and consequently,

$$(A,G) = A^{\Delta(G)}, \quad \Delta_{\mathbb{Z}/p^n\mathbb{Z}}(G) = \Delta(G).$$

Repeating this process we get

$$(A,G,G,\cdots,G) = A^{\Delta(G)^r}.$$

Let $C(A)$ be the centralizer of A in G. Since by hypothesis $G/C(A)$ is a finite p-group its augmentation ideal $\Delta\left(G/C(A)\right)$ is nilpotent (I.2.21). Thus there exists an r for which

$$\Delta^r(G) \subseteq \Delta\left(G,C(A)\right).$$

Therefore,

$$(A,G,G,\cdots,G) \subseteq A^{\Delta\left(G,C(A)\right)}.$$

Since for $c \in C(A)$, $a \in A$, $g \in G$ we have

$$a^{g(1-c)} = (a^g)^{1-c} = (a^g)a^{-g} = 1$$

it follows that $(A,G,G,\cdots,G) = 1$. $\square$

Lemma 4.2. Any p-group G, whose commutator subgroup is finite, is nilpotent.

Proof: We use induction on $|G'|$. Let i be so chosen that

$$A = \gamma_i(G') \neq 1, \quad \gamma_{i+1}(G') = 1.$$

Then A is a finite normal Abelian p-subgroup of G and G acts as a finite group of automorphisms of A. Hence by the last lemma $A \subseteq \zeta_r(G)$, the r-th term of the upper central series of G for a suitable r. Since G/A is nilpotent by induction, it follows that G is nilpotent. □

Lemma 4.3. Let R be a ring of prime characteristic p. Then for $x,y \in R$ we have

$$[x,\underbrace{y,\cdots,y}_{p^m}] = [[\cdots[[x,y],y],\cdots],y] = [x,y^{p^m}].$$

Proof: Let r_y and ℓ_y denote the operators on R which are right multiplication and left multiplication by y. Then

$$[x,y] = xy - yx = (r_y - \ell_y)(x)$$

so that

$$[[\cdots[[x,y],y],\cdots],y] = (r_y - \ell_y)^{p^m}(x).$$

Since right and left multiplication commute as operators and since R has characteristic p we have

$$(r_y - \ell_y)^{p^m}(x) = \left((r_y)^{p^m} - (\ell_y)^{p^m}\right)(x) = xy^{p^m} - y^{p^m}x,$$

proving the result. □

Theorem 4.4 (Passi-Passman-Sehgal [1]). The group algebra KG of the group G over a field K of characteristic $p \geq 0$ is Lie nilpotent if and only if G is p-abelian and nilpotent.

Proof: Suppose that KG is Lie nilpotent; say $(KG)^{[m]} = 0$. Then by (3.4) G is p-abelian. Pick n such that $p^n+1 \geq m$ then $(KG)^{[p^n+1]} = 0$ and by (4.3) for $y \in g$, y^{p^m} commutes with every

$x \in G$. This means that

$$\left(G/\zeta(G)\right)^{p^m} = 1$$

i.e. $\overline{G} = G/\zeta(G)$ is a p-group with finite commutator subgroup. Thus by (4.2) $\overline{G}$ is nilpotent and consequently G is nilpotent.

The converse follows from the next lemma.

Lemma 4.5. Let k be a commutative ring, with identity, having characteristic p^t where p is a prime. Let G be a p-abelian nilpotent group. Then kG is strongly Lie nilpotent.

Proof: We use induction on $|G'|$. If $G' = 1$, G is Abelian and there is nothing to prove. Se we may suppose that $|G'| > 1$.

Let us pick an element z of order p in $G' \cap \zeta(G)$. Write $\overline{G} = G/\langle z\rangle$. Then $\overline{G}$ satisfies the hypothesis of the lemma and $(\overline{G})'$ has smaller order than G'. Consequently there is an m such that

$$(k\overline{G})^{(m)} = 0 \quad \text{i.e.} \quad (kG)^{(m)} \subseteq \Delta(G,\langle z\rangle) = (1-z)kG.$$

Since z is central it follows that

$$kG^{(mp^t)} \subseteq (1-z)^{p^t} kG = 0.$$

This proves the lemma and the theorem. □

Now it is an easy matter to derive criteria for Lie solvability and Lie nilpotence of kG where k is a commutative ring with identity.

Theorem 4.6. Let k be a commutative ring, with identity, of characteristic $\chi \geq 0$. Necessary and sufficient conditions for the Lie solvability of kG are:

(i) G is p-abelian when χ is a power of an odd prime p;

(ii) G is Abelian when χ is 0 or has two distinct odd prime divisors;

(iii) G is p-abelian and has an Abelian subgroup of index at most two when χ is divisible by two and only one odd prime p;

(iv) G has a 2-abelian subgroup of index at most two when χ is a power of 2.

Proof: (i) Let $\chi = p^a$. Then $k/p^{a-1}k = \bar{k}$ has characteristic p and as such $\bar{k}$ contains an isomorphic copy of $\mathbb{Z}/p\mathbb{Z} = K$. Thus kG Lie solvable $\Rightarrow$ KG Lie solvable $\Rightarrow$ G p-abelian by (3.1). The converse is Lemma 3.3.

(ii) Let $\chi = 0$ or $p^a q^b r$ where p and q are distinct odd primes. Assume kG is Lie solvable. Then taking suitable homomorphic images and applying (i) we see that G is p-abelian and q-abelian. It follows that G is Abelian. The converse is trivial.

(iii) Let $\chi = 2^a p^b$ where p is an odd prime. Assume kG is Lie solvable; then $(\mathbb{Z}/p\mathbb{Z})G$ is Lie solvable and G is p-abelian by (3.1). Again since $(\mathbb{Z}/2\mathbb{Z})G$ is Lie solvable, G has a normal 2-abelian subgroup N of index of most 2. Since N is 2-abelian as well as p-abelian it follows that N is Abelian proving the necessity of the conditions. Conversely, suppose that G has a normal Abelian subgroup A of index ≤ 2 and that G is p-abelian. Then it follows by Lemma 3.3 that for suitable n and ℓ

(4.7) $$\delta^{(n)}(kG) \subseteq p^b kG,$$

(4.8) $$\delta^{[\ell]}(kG) \subseteq 2^a kG.$$

Now (4.7) and (4.8) taken together imply that there exists an m such that

$$\delta^{[m]}(kG) \subseteq 2^a p^b kG = 0.$$

(iv) Since $k/2k \supseteq \mathbb{Z}/2\mathbb{Z}$, the necessity part follows from (3.1)(ii). The sufficiency part is (3.3)(ii). □

The same arguments as in the last theorem together with (4.4) and (4.5) give us the next result.

Theorem 4.9. Let k be a commutative ring, with identity, of characteristic $\chi \geq 0$. Necessary and sufficient conditions for the Lie nilpotence of kG are:

(i) G is p-abelian and nilpotent when χ is a power of a prime p;

(ii) G is Abelian when χ is divisible by two distinct primes or is zero.

5. STRONG LIE SOLVABILITY

Theorem 5.1. Let kG be the group ring of G over a commutative ring k and let χ be the characteristic of k. Necessary and sufficient conditions for the strong Lie solvability of kG are:

(i) G is p-abelian if χ is a power of a prime p;

(ii) G is Abelian if χ is divisible by two distinct primes or is zero.

Proof: Let us first suppose that $\chi = p^a$ where p is a prime. Then

$$\delta^{(n)}(kG) = 0 \Rightarrow \delta^{(n)}\left((\mathbb{Z}/p^a\mathbb{Z})G\right) = 0 \Rightarrow \delta^{(n)}\left((\mathbb{Z}/p\mathbb{Z})G\right) = 0$$
$$\Rightarrow \delta^{[n]}\left((\mathbb{Z}/p\mathbb{Z})G\right) = 0.$$

In view of Theorem 3.1 and Lemma 3.3, in order to complete the proof of (i), it remains to prove

(5.2) $\delta^{(n)}(KG) = 0,\ K = \mathbb{Z}/2\mathbb{Z} \Rightarrow G$ is 2-abelian.

Assume that $\delta^{(n)}(KG) = 0$. Then by (3.1) we have

$$\left(G\colon \phi(G)\right) \leq 2, \quad |\phi(G)'| < \infty.$$

Suppose that $\left(G\colon \phi(G)\right) = 2$. Then replacing G by $\overline{G} = G/\phi(G)'$ we have

$$(G\colon A) = 2, \quad A = \phi(G), \quad A' = \{1\}.$$

Now in the notation of §2 we have

$$R = (KA)\rho(KG) \supseteq M_2(S)$$

where $S = \Delta\big((A,x_2)\big)$ is not nilpotent. But strong Lie solvability of KG implies that R and thus $M_2(S)$ is strongly Lie solvable. But this is a contradiction as

$$\delta^{(n)}\big(M_2(S)\big) \supseteq M_2(S^{n+1}) \neq 0.$$

This can be seen by observing that for $s,s' \in S$,

$$[se_{12}, s'e_{21}] = ss'(e_{11}+e_{22}).$$

This implies that $\big(G: \phi(G)\big) = 1$ proving (5.2) and hence (i).

(ii) Let $\chi = 0$ or pqr, p, q distinct primes. Then for distinct p, q:

$$\delta^{(n)}(kG) = 0 \Rightarrow \delta^{(n)}\big((\mathbb{Z}/p\mathbb{Z})G\big) = 0,\ \delta^{(n)}\big((\mathbb{Z}/q\mathbb{Z})G\big) = 0$$

and thus G is p-abelian as well as q-abelian. Hence G is Abelian. Conversely if G is Abelian, certainly, $\delta^{(1)}(kG) = 0$. □

The following theorem for finite groups is due to Sandling [2].

Theorem 5.3. Let kG be the group ring of G over a commutative ring k with 1 of characteristic $\chi \geq 0$. Necessary and sufficient conditions for strong Lie nilpotence of kG are:

(i) G is p-abelian and nilpotent if χ is a power of a prime p;

(ii) G is Abelian if χ is divisible by two distinct primes or is zero.

Proof: (4.9) and (4.5).

6. LIE n-ENGEL GROUP RINGS

We wish to characterize group rings KG which are Lie n-Engel. The main result is

Theorem 6.1. Let K be a field with char $K = p \geq 0$. Necessary and sufficient conditions for KG to be Lie n-Engel are:

(i) G is nilpotent and contains a normal p-abelian subgroup A with G/A a finite p-group if $p > 0$;

(ii) G is Abelian if K has characteristic 0.

We need the following lemmas.

Lemma 6.2. Let A be an Abelian normal subgroup of a nilpotent group G. Then

$$(A,G^n) = 1 \Rightarrow (A,G)^{n^e} = 1$$

where e is the class of nilpotence of G.

Proof: Using the identity

$$(a,gh) = (a,h)(a,g)^h = (a,h)(a,g)(a,g,h)$$

for $a \in A$, $g,h \in G$ we obtain

$$(a,g^n) \equiv (a,g)^n \bmod (A,G,G).$$

Consequently

$$(A,G)^n \subseteq (A,G,G). \tag{6.3}$$

Assume that

$$(A,G)^{n^{i-1}} \subseteq (A,{}_iG)$$

then

$$(A,G)^{n^i} \subseteq (A,{}_iG)^n \subseteq (A,{}_{i+1}G).$$

The last containment follows by applying (6.3) with A replaced by $A_1 = (A,{}_{i-1}G)$ which is justified since the hypothesis is satisfied by A_1. It follows now that

$$(A,G)^{n^e} \subseteq (A,{}_{e+1}G) = \{1\},$$

Lemma 6.4. Let R be a commutative ring generated by $a_1,a_2,\cdots,a_m$ such that $a_i^n = 0$ for $1 \le i \le m$. Then $R^{mn} = 0$.

Proof: We have to prove that $\alpha_1\alpha_2\cdots\alpha_{mn} = 0$ for every choice of $\alpha_i \in R$. Clearly each α_k is of the form

$$\sum a_1^{i_1} a_2^{i_2} \cdots a_m^{i_m} .$$

Multiplying together and using the distributive law, each term in the expansion of $\alpha = \alpha_1\alpha_2\cdots\alpha_{mn}$ is a product of at least mn a's. Since we have only m distinct a's at least one of the a's appears at least n times in each product, making it zero. Thus $\alpha = 0$. □

6.5 Proof of (ii): Assume that K has characteristic 0 and KG is Lie n-Engel. Picking a prime p,

$$KG \text{ Lie n-Engel} \Rightarrow \mathbb{Z}G \text{ Lie n-Engel} \Rightarrow \mathbb{Z}/p\mathbb{Z}G \text{ Lie n-Engel}.$$

Choose m such that $p^m \geq n$ then by Lemma 4.3

$$x,y \in G \Rightarrow [x^{p^m},y] = 0$$

and hence $G/\zeta(G)$ is a p-group. Since p is arbitrarily chosen, $G = \zeta(G)$ is Abelian. The converse being trivial, the proof of 6.1(ii) is complete.

6.6 Proof of the necessity part of (i): Assume that K has characteristic $p > 0$. By Theorem 1.1 we have

$$\big(G: \phi(G)\big) < \infty, \quad |\phi(G)'| < \infty.$$

Also by Lemma 4.3 there exists an m such that $G^{p^m} \subseteq \zeta(G)$. Since $\zeta(G) \subseteq \phi(G)$ it follows that $G/\phi(G)$ is a finite p-group. In order to prove that $\phi(G)'$ is a p-group we may assume that $\phi(G)$ is finitely generated. Then $\big(\phi(G): \zeta(\phi(G))\big) < \infty$ and it follows by (I.4.2) that $\phi(G)'$ is a p-group.

It remains to prove that G is nilpotent. First observe that G acts as a finite p-group of automorphisms of $\phi(G)'$. By applying Lemma 4.1 to $G/\phi(G)''$ with $A = \phi(G)'/\phi(G)''$ we obtain

$$\big(\phi(G)',G,G,\cdots,G\big) \subseteq \phi(G)''.$$

Repeating this process we obtain that for a suitable r, $(\phi(G)',{}_rG) = 1$. Thus, in order to prove the nilpotence of G, we may assume that $\phi(G)' = 1$. In other words, we have a normal Abelian subgroup A of G such that G/A is a finite p-group and $G^{p^m} \subseteq \zeta(G)$. Again, we may factor by the central subgroup A^{p^m} and assume that $A^{p^m} = 1$. Thus by Lemma 4.1, there exists an r such that

$$(A,{}_rG) = 1, \quad A \subseteq \zeta_r(G).$$

Hence G is nilpotent. □

The next statement completes the proof of (i) and hence the theorem.

(6.7) Let G be a nilpotent group of class e with a normal subgroup A with both G/A and A' finite p-groups. If k is a commutative ring, with identity, of characteristic p then kG is Lie n-Engel for some n.

Proof of (6.7): We use induction on $|A'|$. If $|A'| > 1$ we choose an element z of order p in $A' \cap \zeta(G)$. Then

$$G/\langle z\rangle \rhd A/\langle z\rangle \rhd A'/\langle z\rangle$$

and therefore by induction and (4.3) for all γ, $\mu \in kG$ there exists a suitable ν such that

$$[\gamma,\mu^{p^\nu}] \in (1-z)kG = \Delta(G,\langle z\rangle).$$

Thus,

$$[\gamma,\underbrace{\mu^{p^\nu},\cdots,\mu^{p^\nu}}_{p}] \in (1-z)^p kG = 0.$$

It follows by Lemma 4.3 that $[\gamma,\mu^{p^{\nu+1}}] = 0$. Hence it remains to prove (6.7) when

(6.8) A is Abelian.

We assert that for a suitable m

$$G^{p^m} \subseteq \zeta(G). \tag{6.9}$$

This can be seen as follows. By the hypothesis and (6.8) there exists an s such that $(A, G^{p^s}) = 1$ and thus by (6.2) $(A,G)^{p^{es}} = 1$ which implies that $(A^{p^{es}}, G) = 1$. Thus $G^{p^m} \subseteq \zeta(G)$ for a suitable m.

We have the situation:

$$G \rhd A, \quad A' = \{1\}, \quad G/A \text{ finite p-group}, \quad G^{p^m} \subseteq \zeta(G).$$

Let $T = \{1 = t_1, t_2, \cdots, t_n\}$ be a set of coset representatives of A in G. Then every element ξ of kG can be represented uniquely as

$$\xi = \sum_{t \in T} \alpha_t t, \quad \alpha_t \in kA.$$

Therefore,

$$\xi^{p^m} = \sum \alpha_t^{p^m} t^{p^m} + \eta, \quad \eta \in \Delta(G,G')$$
$$= \theta + \eta$$

where $\theta = \Sigma \alpha_t^{p^m} t^{p^m}$ is an element of the center of kG. Writing $B = (A,G)$ we observe that G'/B is a finite p-group in view of the fact that $A/B \subseteq \zeta(G/B)$ and that G/A is a finite p-group (see (I.4.2)). We therefore have an integer ℓ such that

$$\left(\Delta(G')\right)^{p^\ell} \subseteq \Delta(B)kG.$$

Consequently,

$$(\xi^{p^m})^{p^\ell} = \theta^{p^\ell} + \eta^{p^\ell} = \theta^{p^\ell} + \delta, \quad \delta \in \Delta(G,B). \tag{6.10}$$

Let us write $\delta = \Sigma_{t \in T} \delta_t t$ with $\delta_t \in \Delta(A,B)$. Let I_δ be the ring generated by all $\delta_t^\tau = \tau^{-1}\delta_t \tau$, $t \in T$, $\tau \in T$. Then $I_\delta \subseteq \Delta(A,B)$ is commutative. Moreover, since $(G^{p^m}, A) = 1$, it follows by (6.2) that $B^{p^r} = 1$ for a fixed r. Consequently,

$$\delta_t^{p^r} = 0 \text{ for all } t. \tag{6.11}$$

Thus I_δ is generated by $\leq |T|^{|T|}$ commuting elements each satisfying (6.11). It follows by (6.4) that

$$I_\delta^w = 0 \quad \text{where} \quad w = p^r \cdot |T|^{|T|}.$$

Thus $I_\delta^{p^M} = 0$ where M is independent of δ but depends on $|T|$ and r only. Thus $\delta^{p^M} = 0$ and (6.10) implies that

$$\xi^{p^{M+m+\ell}} \in \text{center of } kG.$$

Hence kG is Lie n-Engel for $n = p^{M+m+\ell}$. □

The next theorem has been proved by Sandling [2] for finite groups.

<u>Theorem 6.12</u>. Let k be a commutative ring, with identity, of characteristic χ. Necessary and sufficient conditions for kG to be n-Engel are:

(i) G is nilpotent and contains a normal p-abelian subgroup A with G/A a finite p-group if χ is a power of a prime p;

(ii) G is Abelian if χ is divisible by two distinct primes or is zero.

<u>Proof</u>: (i) Assume that $\chi = p^a$. Then k/pk has characteristic p and contains an isomorphic copy of $\mathbb{Z}/p\mathbb{Z}$. Thus

$$kG \text{ Lie n-Engel} \Rightarrow \mathbb{Z}/p\mathbb{Z}G \text{ Lie n-Engel}.$$

The necessity of the condition follows by Theorem 6.1. Conversely, assume that G satisfies the conditions spelled out in the statement of (i) then by (6.7) k/pkG is Lie n-Engel i.e. for all γ, $\mu \in kG$

$$[\gamma, \mu, \cdots, \mu] \in pkG.$$

It is easy to see that kG is $p^{a-1}n$-Engel.

(ii) If χ is zero or divisible by two distinct primes then there exist distinct primes p, q such that $\mathbb{Z}/p\mathbb{Z}(G)$ and $(\mathbb{Z}/q\mathbb{Z})G$ are Lie n-Engel. Hence by (6.2), $G/\zeta(G)$ is of p-power exponent as well as of q-power exponent. Thus $G = \zeta(G)$ is Abelian. And conversely if G is Abelian, kG is clearly 1-Engel. □

We close this chapter with a few remarks. It may be observed that the following statements are equivalent if K is a field of characteristic $p > 0$.

(6.13) $(KG)^{[n]} = 0$ for some n;

(6.14) G is nilpotent p-abelian;

(6.15) $(KG)^{(m)} = 0$ for some m.

The implication (6.13) ⟹ (6.14) is Theorem 4.4 whereas (6.14) ⟹ (6.15) is (4.5). Moreover, trivially, (6.15) implies (6.13).

Let $\mathcal{U} = \mathcal{U}(KG)$. For $x,y \in \mathcal{U}$ we have

$$\begin{aligned}(x,y) - 1 &= x^{-1}y^{-1}xy - 1 \\ &= x^{-1}y^{-1}(xy-yx) \\ &= x^{-1}y^{-1}[x,y].\end{aligned}$$

This together with the equation

$$(ab-1) = (a-1) + (b-1) + (a-1)(b-1)$$

implies, by induction on n, that

$$\gamma_n(\mathcal{U}) - 1 \subseteq (KG)^{(n)}, \quad \delta_n(\mathcal{U}) - 1 \subseteq \delta^{(n)}(KG).$$

Hence we have proved:

(6.16) KG strongly Lie nilpotent $\Longleftrightarrow$ G nilpotent p-abelian $\Rightarrow$ $\mathcal{U}$ nilpotent.

Also, we have seen that if $p \neq 2$, then

(6.17) KG Lie solvable $\Longleftrightarrow$ KG strongly Lie solvable
$\Longleftrightarrow$ G p-abelian $\Rightarrow$ $\mathcal{U}$ solvable.

(6.18) If $p = 2$ then KS_3 is Lie solvable but not strongly Lie solvable.

CHAPTER VI

UNITS IN GROUP RINGS II

In this chapter, we shall characterize groups G such that the unit group, $\mathcal{U}KG$, of the group ring KG satisfies certain nice preassigned properties. Some of these properties have already been considered in Chapter II. We shall need to know if in certain cases KG has (nonzero) nilpotent elements.

1. NILPOTENT ELEMENTS

Recall that a unit of KG is said to be trivial if it is of the form ag, $a \in K$, $g \in G$. We have seen (II.4.1) that, in general, the group ring of a finite group contains nontrivial units. In contrast if $\langle x \rangle$ is an infinite cyclic group and K is a field then every unit of $K\langle x \rangle$ is trivial. This can be seen by letting γ be a unit with inverse μ and considering the highest and the lowest degree terms in the product $\gamma\mu$. The same argument gives the fact that all units of the group algebra of an ordered group over a field are trivial (see (1.6)).

A group is called an up-group (unique product group) if given any two nonempty subsets A and B of G there exists at least one element $x \in G$ which can be represented uniquely as $x = ab$, $a \in A$, $b \in B$. For these groups there is the

Theorem 1.1 (Passman-Sehgal). If G is an up-group and K is a field of characteristic zero then all units of KG are trivial.

Proof: (See Passmann [6], p. 590).

In fact there is the famous

Conjecture 1.2. All units of the group algebra of a torsion free group over a field are trivial.

In addition there is an apparently weaker conjecture as follows.

Conjecture 1.3. The group algebra of a torsion free group over a field has no zero divisors.

That this conjecture follows from the previous one is seen below.

Proposition 1.4. Let K be a field and G a torsion free group.

(i) KG has zero divisors if and only if it has (nonzero) nilpotent elements.

(ii) All units of KG are trivial $\Rightarrow$ KG has no zero divisors.

Proof: (i) Clearly, if KG has a (nonzero) nilpotent element then it has a zero divisor. Suppose there exist elements $a \neq 0$, $b \neq 0$ of KG such that $ab = 0$. Then since KG is a prime ring (see Passman [6]) $b(KG)a \neq 0$ but $\left(b(KG)a\right)^2 = 0$. Thus we have a nonzero element $\gamma \in b(KG)a$ with $\gamma^2 = 0$.

(ii) Suppose that KG has a zero divisor. Then by (i) it has a nonzero element γ with $\gamma^2 = 0$. Therefore, $(\gamma+1)(\gamma-1) = 1$ and $\gamma + 1 \in \mathcal{U}KG$. Thus $(\gamma+1) = kg$, $k \in K$, $g \in G$. But since the augmentation of γ is zero we can conclude that $k = 1$. We have

$$\gamma = (g-1), \quad \gamma^2 = 0$$

which is clearly false since g has infinite order. □

We shall find it useful to know that all units of the crossed product of an ordered group over a field are trivial. A group G is said to be ordered if there exists an order relation, $<$, on G

such that any two elements $g_1, g_2 \in G$ can be compared and $g_1 < g_2$ implies $ag_1 < ag_2$ and $g_1 a < g_2 a$ for all $a \in G$. The crossed product $K(G,\rho,\sigma)$ is defined as follows.

Definition. Let G be a group and K a ring with identity. Suppose that we are given a function $\rho: G \times G \to \mathcal{U}K$, called a factor system; and automorphisms σ_g of K for each $g \in G$. Suppose that ρ and σ satisfy the following properties for each $g,h,\ell \in G$, $a \in K$:

(i) $$\rho_{g,h}\rho_{gh,\ell} = \sigma_g(\rho_{h,\ell})\rho_{g,h\ell}$$

and

(ii) $$\rho_{h,g}\sigma_{hg}(a) = \sigma_h\big(\sigma_g(a)\big)\rho_{h,g} .$$

Then by the crossed product $K(G,\rho,\sigma)$, of G over K with factor system ρ and automorphisms σ, we understand the set of finite sums

$$\left\{\sum a_i \bar{g}_i : a_i \in K, \quad g_i \in G\right\},$$

where $\bar{g}_i$ is a symbol corresponding to g_i. Equality and addition are defined componentwise and for $g,h \in G$, $a \in K$ we have

$$\bar{g}\,\bar{h} = \rho_{g,h}\overline{gh}\ , \quad \bar{g}\,a = \sigma_g(a)\bar{g}\ . \qquad (*)$$

$K(G,\rho,\sigma)$ is easily seen to be a ring if we extend (*) distributively.

As a special case, if $\rho_{g,h} = 1$ for all $g,h \in G$ we get the skew group ring $K(G,1,\sigma)$ and we denote it by $K_\sigma(G)$. In addition, if we have $\sigma_g = I$ for all $g \in G$ we get the group ring KG. The crossed products arise naturally. For example, if G has a normal subgroup N then picking a fixed set of coset representatives $\{\bar{g}_v\}$ of N in G we can write every $\alpha \in KG$ as $\alpha = \Sigma a_g \bar{g}$, $a_g \in KN$. Then for $g,h \in G$, $a \in KN$, we write

$$\bar{g}\,\bar{h} = \rho_{g,h}\overline{gh}\ , \quad \bar{g}\,a = \sigma_g(a)\bar{g}\ .$$

It follows that

(1.5) $$KG = (KN)(G/N,\rho,\sigma).$$

Proposition 1.6. Suppose that G is an ordered group and K is a ring without zero divisors. Then the units of any crossed product $K(G,\rho,\sigma)$ are trivial. Also, $K(G,\rho,\sigma)$ has no zero divisors.

Proof: Suppose that we have $\gamma,\mu \in K(G,\rho,\sigma)$ with $\gamma\mu = 1$. We may write

$$\gamma = \gamma_1\overline{g}_1 + \gamma_2\overline{g}_2 + \cdots + \gamma_s\overline{g}_s , \quad g_1 < g_2 < \cdots < g_s$$

$$\mu = \mu_1\overline{h}_1 + \mu_2\overline{h}_2 + \cdots + \mu_t\overline{h}_t , \quad h_1 < h_2 < \cdots < h_t$$

with γ_i, μ_j nonzero elements of K. We have to prove that $s = 1$. Suppose to the contrary. Then among the products $\{g_ih_j, 1 \le i \le s, 1 \le j \le t\}$, g_1h_1 is the smallest and g_sh_t is the largest. Now,

$$\begin{aligned} 1 = \gamma\mu &= \gamma_1\sigma_{g_1}(\mu_1)\rho_{g_1,h_1}\overline{g_1h_1} + \cdots + \gamma_s\sigma_{g_s}(\mu_t)\rho_{g_s,h_t}\overline{g_sh_t} \\ &= c_1\overline{g_1h_1} + \cdots + c_m\overline{g_sh_t} , \quad c_1,c_m \neq 0. \end{aligned}$$

It follows that $g_1h_1 = 1 = g_sh_t$ and

$$g_1 < g_s \Rightarrow g_1^{-1} > g_s^{-1} \Rightarrow h_1 > h_t$$

which is a contradiction, proving the triviality of units. The argument also proves that $K(G,\rho,\sigma)$ has no zero divisors. □

Corollary 1.7. Let G be torsion free nilpotent and K a ring without zero divisors. Then the units of any crossed product $K(G,\rho,\sigma)$ are trivial. Also, $K(G,\rho,\sigma)$ has no zero divisors.

Proof: First we may assume that G is finitely generated. We have only to see that G can be ordered. We know (I.3.8) that the upper central series $\zeta_i(G)$ of G has torsion free finitely generated and hence ordered factors. We may order G lexicographically as follows: If $\zeta_i(G) = \zeta_i$ is already ordered we express every element g of $\zeta_{i+1}(G) = \zeta_{i+1}$ as $g = xy$, $x \in T$, a set of coset representatives of ζ_{i+1}/ζ_i, $y \in \zeta_i$. For $g_1 = x_1y_1$, $g_2 = x_2y_2$, $x_1,x_2 \in T$, $y_1,y_2 \in \zeta_i$, set

$$g_1 < g_2 \iff \text{either } x_1\zeta_i < x_2\zeta_i \text{ or } x_1\zeta_i = x_2\zeta_i, \ y_1 < y_2.$$

One can check that this is an order on ζ_{i+1}. Thus, by induction, G is ordered and the result follows by the last proposition. □

The argument in (1.6) also proves the next result.

Proposition 1.8. Let $R(G,\rho,\sigma)$ be the crossed product of an ordered group over a ring R such that for any $a \in R$, $g \in G$:

$$a\sigma_g(a) = 0 \Rightarrow a = 0.$$

Then $R(G,\rho,\sigma)$ has no (nonzero) nilpotent elements.

Now we are ready to describe group rings which have no nonzero nilpotent elements. The case of finite groups was settled independently by Pascaud [1] and Sehgal [9] who also answered the question for nilpotent and FC-groups. The main theorem gives necessary and sufficient conditions for QG to have no nilpotent elements if the crossed product $R(\overline{G},\rho,\sigma)$ has no nilpotent elements for a certain torsion free homomorphic image $\overline{G}$ of G. We shall say a ring "R has no nilpotent elements" if

$$r \in R, \quad r^2 = 0 \Rightarrow r = 0.$$

Lemma 1.9. Every idempotent e of a ring R, without nilpotent elements, is central.

Proof: Since $\big(eR(1-e)\big)^2 = \big((1-e)Re\big)^2 = 0$ we have

$$er = ere = re \quad \text{for all } r \in R. \ \square$$

Lemma 1.10. Let G be a finite group and K a commutative ring. Suppose that $\gamma = \Sigma\gamma(g)g \in KG$ is nilpotent. Then $|G|\gamma(1)$ is a nilpotent element of R.

Proof: Let P be a prime ideal of K. Look at the image of γ in $(K/P)G$. Considering the regular representation matrix of γ and taking trace we get that

$$|G|\gamma(1) \equiv 0 \pmod{P}.$$

Thus $|G|\gamma(1)$ belongs to all prime ideals P of R and hence is nilpotent. □

Lemma 1.11. If the group algebra KG of G over a field K has no nilpotent elements than the torsion elements, $T(G)$, of G form an Abelian or a Hamiltonian group.

Proof: If K has characteristic $p > 0$ then G has no element of order p as $(g-1)^p = 0$. Let H be a finite subgroup of G. Then

$$e = \frac{1}{|H|} \sum_{h \in H} h = e^2$$

is central by (1.9). Thus H is normal in G. Hence the torsion elements of G from a normal subgroup which is Abelian or Hamiltonian

Proposition 1.12 (Pascaud [1], Sehgal [9]). Let G be finite and K a commutative ring of characteristic $n > 0$. Then KG has no nilpotent elements if and only if $|G|$ is not a zero divisor in K and G is Abelian.

Proof: We shall first prove the necessity of the conditions. Suppose that $|G|r = 0$ for some $r \in R$. Then $r\Sigma_{g \in G}\, g = \gamma$ satisfies $\gamma^2 = 0$. Thus $\gamma = 0$ which implies that $r = 0$. Now, to prove the commutativity of G we observe that n is square free and so $n = p_1 p_2 \cdots p_k$, a product of distinct primes.

$$KG \supseteq (\mathbb{Z}/n\mathbb{Z})G = (\mathbb{Z}/p_1\mathbb{Z})G \oplus \cdots \oplus (\mathbb{Z}/p_k\mathbb{Z})G .$$

Since $(\mathbb{Z}/p_1\mathbb{Z})G$ has no nilpotent elements, G has no element of order p_1. Thus by Wedderburn's theorem

$$(\mathbb{Z}/p_1\mathbb{Z})G \simeq \sum^{\oplus} (D_i)_{n_i} ,$$

a direct sum of full matrix rings over division rings D_i. Since each D_i is finite it is commutative. Also by (1.9) or otherwise each $n_i = 1$. We have proved that G is commutative.

In order to prove the converse, suppose that

$$0 \neq \gamma = \sum \gamma(g) g \in KG$$

is nilpotent. We may suppose, by considering γg^{-1} if necessary for some g, that $\gamma(1) \neq 0$. It follows by (1.10) that $|G|\gamma(1)$ is nilpotent and therefore by hypothesis $|G|\gamma(1) = 0$. Hence $\gamma(1) = 0$ which is a contradiction proving the proposition. □

We recall that a group G is Hamiltonian if and only if

$$G = A \times E \times K_8$$

where A is an Abelian group in which every element has odd order, E is an elementary Abelian 2-group, i.e. $E^2 = 1$ and K_8 is the quaternion group of order 8 given by

$$K_8 = \langle i,j: i^2 = j^2 = t, \quad t^2 = 1, \quad ji = ijt \rangle .$$

We need to know when RK_8 can have nilpotent elements.

Proposition 1.13. Let R be a commutative ring which has no nilpotent elements. Suppose that 2 is not a zero divisor in R. Then RK_8 has no nilpotent elements if and only if the equation

$$a^2 + b^2 + c^2 = 0, \quad a,b,c \in R$$

has no nonzero solution.

Proof: Let $\gamma = \Sigma\gamma(g)g \in RK_8$ be such that $\gamma^2 = 0$. Since $G/\langle t \rangle$ is Abelian, by the last result, we have

$$\gamma \in \Delta_R(K_8,\langle t \rangle).$$

Also, we can conclude from (1.10) that $\gamma(1)$ and $\gamma(t)$ are both zero. Thus γ is of the form

$$\gamma = (ai+bj+cij)(1-t), \quad a,b,c \in R.$$

Then

$$\gamma^2 = 2(a^2+b^2+c^2)(t-1) = 0 \iff a^2 + b^2 + c^2 = 0.$$

The proposition is proved. □

It follows that QK_8 has no nilpotent elements. As a matter of fact, we have

<u>Proposition 1.14</u>. $QK_8 \simeq Q \oplus Q \oplus Q \oplus Q \oplus S$ where S is the rational quaternion algebra $\{Q \dotplus QI \dotplus QJ \dotplus QK\}$. Moreover, $\mathcal{U}QK_8$ is not solvable.

<u>Proof</u>: Consider the homomorphism

$$QK_8 \overset{\sigma}{\to} S, \quad \sigma(i) = I, \quad \sigma(j) = J$$

Kernel of σ is $(1+t)QK_8$. We have

$$QK_8 = (1+t)QK_8 \oplus (1-t)QK_8$$

as $\frac{1}{2}(t+1)$ is a central idempotent. Since $(1+t)QK_8$ is the kernel of σ it follows that σ is 1-1 on $(1-t)QK_8$. Now,

$$S = \sigma(QK_8) = \sigma\big((1-t)QK_8\big) \simeq (1-t)QK_8$$

and

$$(1+t)QK_8 \simeq Q(K_8/\langle t\rangle) \simeq Q \oplus Q \oplus Q \oplus Q$$

by (II.2.6) as $K_8/\langle t\rangle$ is the Klein four group. we get

$$QK_8 \simeq Q \oplus Q \oplus Q \oplus Q \oplus S.$$

Notice that the multiplicative group $\dot{S}$ of S is not solvable. This may be seen by the fact that $\dot{S}/\dot{Q}$ contains no non-identity Abelian normal subgroup. It follows that $\mathcal{U}QK_8$ is not solvable. □

We have seen that if G is finite non-commutative and QG has no nilpotent elements then $G = A \times E \times K_8$. Therefore

$$QG = \big(Q(A \times E)\big)K_8.$$

Thus it remains to decide when RK_8 has nilpotent elements for cyclotomic fields R. In view of Proposition 1.13 the next theorem is just the required result.

Theorem 1.15 (C. Moser). Let $Q_m = Q(\xi_m)$ be a cyclotomic field where ξ_m is a primitive m-th root of unity and $m > 1$ is an odd integer. Then the equation

$$-1 = a^2 + b^2$$

has a solution $x,y \in Q_m$ if and only if the multiplicative order of 2 modulo m is even.

Proof: See Moser [1].

Let A be a finite Abelian group. The QA has d primitive idempotents where d is the number of cyclic factor groups of A (II.2.6). On the other hand, there are the idempotents

$$\varepsilon_C = \frac{1}{|C|} \sum_{x \in C} x$$

for any subgroup C of A. Every idempotent of QA is actually a Q-linear combination of the ε_C's.

Proposition 1.16. Every idempotent of QA is a Q-linear combination of the ε_C's.

Proof: It suffices to prove that every primitive idempotent of QA is a linear combination of the ε_C's. Consider the set $\{\varepsilon_C\colon A/C \text{ cyclic}\}$. This set has the same number of elements as the total number of primitive idempotents of QA. So it is enough to prove that this set is linearly independent over Q. If not, suppose

$$\sum \alpha_i \varepsilon_{C_i} = 0 \qquad 0 \neq \alpha_i \in Q.$$

Let C_0 be a minimal group among the C_i's. Let χ be the complex character of A with kernel C_0. Then since ε_C is an idempotent, $\chi(\varepsilon_C) = 0$ or 1. On the other hand,

$$|\chi(\varepsilon_C)| = \left| \frac{1}{|C|} \sum_{x \in C} \chi(x) \right| \leq 1$$

Thus
$$\begin{aligned} \chi(\varepsilon_C) = 1 &\iff \chi(x) = \chi(x') \quad \text{for all } x,x' \in C \\ &\iff \chi(x) = 1 \quad \text{for all } x \in C \\ &\iff C \subseteq \operatorname{Ker} \chi \iff C = C_0. \end{aligned}$$

Applying χ to the dependency relation we get $\alpha_0 = 0$ which is a contradiction, proving the result. □

The next theorem is announced in Berman [1] and proved in Pascaud [1] and Sehgal [9].

Theorem 1.17. The rational group algebra QG of a finite group G has no nilpotent elements if and only if one of the following is satisfied.

(i) G is Abelian,

(ii) G is Hamiltonian of order $2^n m$, m odd, such that the multiplicative order of $2 \bmod m$ is odd.

Proof: Suppose that G is a non Abelian group such that QG has no nilpotent elements. Then by (1.11) and (1.9)

$$\begin{aligned} G &= A \times E \times K_8, \\ QG &= (QE)(A \times K_8) = \left(\sum^{\oplus} Q\right)(A \times K_8), \\ &= \sum^{\oplus} Q(A \times K_8) = \sum^{\oplus} (QA)K_8. \end{aligned}$$

It is enough to consider $(QA)K_8$. But we know (II.2.6) that

$$QA = \sum_{d|e}^{\oplus} n_d Q_d$$

where $n_d Q_d$ denotes n_d copies of $Q(\xi_d)$, ξ_d is a primitive d-th root of unity and e is the exponent of A. Then

$$(QA)K_8 = \sum_{d|e} n_d Q_d(K_8).$$

Now (ii) follows from (1.13) and (1.15). The same argument together with (1.13) and (1.15) give the converse. □

We make a couple of useful observations whose proofs are easy.

Lemma 1.18. Let B be a finite Abelian subgroup of G. Then every subgroup of B is normal in G if and only if every $x \in G$ induces an automorphism

$$B \ni b \to x^{-1}bx = b^i, \quad i = i(x)$$

of B.

Lemma 1.19. Let K_8 be a normal subgroup of G. Then every subgroup of K_8 is normal in G if and only if every $x \in G$ induces on K_8, one of its four inner automorphisms.

Theorem 1.20 (Sehgal [9]). Suppose that QG has no nilpotent elements. Then

(1.21) T = T(G), the set of torsion elements of G, forms a normal subgroup of G,

and one of the following is satisfied.

(1.22) T is Abelian and for $x \in G$, we have locally on T

$$x^{-1}tx = t^i \quad \text{for all} \quad t \in T, \quad i = i(x)$$

i.e. for a finite subgroup B of T we have $x^{-1}bx = b^i$ for all $b \in B$ and i depends on B and x.

(1.22') $T(G) = A \times E \times K_8$ where A is an Abelian group in which every element has odd order, E is an elementary Abelian 2-group and K_8 is the quaternion group such that modulo every n, with an element of order n in A, the multiplicative order of 2 is odd. Moreover,

(1) K_8 is normal in G,

(2) Conjugation by $x \in G$ induces an inner automorphism on K_8 and

(3) Conjugation by $x \in G$ acts as in (1.22) on $A \times E$.

Conversely, suppose that G is any group satisfying (1.21) and (1.22) or (1.22'). Further suppose

(1.23) Any crossed product $R(G/T,\rho,\sigma)$ which has the property:

$$a\sigma_g(a) = 0 \quad \text{for} \quad a \in R, \quad g \in G \Rightarrow a = 0,$$

has no nilpotent elements.

Then QG has no nilpotent elements.

Proof: Suppose that QG has no nilpotent elements. Then since every idempotent of QG is central by (1.9), it follows that every finite subgroup of G is normal in G. Hence T(G) is a group which is either Abelian or $T(G) = A \times E \times K_8$. If T(G) is Abelian, (1.22) follows from (1.18). Now, we have to consider the case that $T(G) = A \times E \times K_8$. Let us suppose that A has an element g of prime order p. Then

$$Q(\langle g\rangle \times K_8) = \big(Q \oplus Q(\xi_p)\big)K_8 = K_8 \oplus Q(\xi_p)K_8.$$

It follows by (1.13) and (1.15) that the multiplicative order of 2 modulo p is odd. It is easily seen that order of 2 is odd modulo every n for which there is an element of order n in A. Thus (1.18) and (1.19) complete the proof of the necessity part of the theorem.

To prove the sufficiency part of the theorem we observe that

$$QG = (QT)(G/T,\rho,\sigma).$$

In view of (1.23) it is enough to prove

$$a\sigma_x(a) = 0 \quad \text{for} \quad a \in QT, \quad \overline{x} = g \in G/T \Rightarrow a = 0.$$

Clearly $a \in QT_0$ where T_0 is finite. Either T_0 is Abelian in which case

$$QT_0 = F_1 \oplus \cdots \oplus F_k,$$

a direct sum of fields. Then every idempotent of QT_0 is central in QG due to (1.22) as seen by (1.16). Again σ_x maps QT_0 onto QT_0 and each F_i onto itself. It follows that $a\sigma_x(a) = 0 \Rightarrow a = 0$.

Now we have to consider the case that $T_0 = A \times E \times K_8$ where A and E are finite Abelian, $E^2 = 1$. We wish to prove that every idempotent of QT_0 is central in QG. Since by Theorem 1.17 QT_0 has no nilpotent elements, it follows that every idempotent of QT_0 is central in QT_0. Writing

$$A_1 = A \times E, \quad T_0 = A_1 \times K_8$$

we have

$$QA_1 = K_1 \oplus \cdots \oplus K_\ell ,$$

a direct sum of fields. Moreover, every idempotent of QA_1 is central in QG by (1.16). Also, as proved in (1.14) we have

$$QK_8 = Q \oplus Q \oplus Q \oplus Q \oplus S$$

where S is the skew field of the rational quaternions. If we pick an $x \in G$, $x \notin T$ then it can be checked directly that $Q\langle K_8,x\rangle$ has no nilpotent elements by using the fact that x induces one of the four inner automorphisms on K_8. Suppose that we have a $\gamma \in Q\langle K_8,x\rangle$ such that $\gamma^2 = 0$. Notice that $H = \langle K_8,x\rangle \supset K_8 \supset \langle t\rangle$ where the torsion subgroup of $H/\langle t\rangle$ is Abelian. Since $H/\langle t\rangle$ satisfies the required condition we have by (1.8), $\overline{\gamma} = 0$ in $Q(H/\langle t\rangle)$ This implies that $\gamma \in \Delta(H,\langle t\rangle)$. Thus γ is of the form

$$\gamma = x^n(a_0+a_1i+a_2j+a_3ij)(1-t) + \cdots$$

where x^n is the highest degree nonzero term in x appearing in the support of γ. Then supposing $x^{-n}ix^n = i$, $x^{-n}jx^n = jt$, we get

$$\begin{aligned}\gamma^2 &= 2x^{2n}(a_0+a_1i+a_2jt+a_3ijt)(a_0+a_1i+a_2j+a_3i)(1-t) + \cdots \\ &= 2x^{2n}(a_0^2+a_2^2+a_3^2+a_1^2t+2a_1a_0+2a_1a_2j+a_2a_3ij+a_1a_3jt)(1-t) + \cdots \\ &= 0.\end{aligned}$$

It is now easy to see that $\gamma = 0$. We reach the same conclusion similarly if x^n inducing one of the other inner automorphisms on

K_8. We conclude by (1.9) that every idempotent of QK_8 is central in QG, in particular, $x^{-1}Sx = S$ for all $x \in G$. Now, we have

$$Q(A_1 \times K_8) = \sum K_i \otimes Q \oplus \sum K_i \otimes Q \oplus \sum K_i \otimes Q \oplus \sum K_i \otimes Q \oplus \sum K_i \otimes S.$$

Each $K_i \otimes S$ is a simple ring (Herstein [2], page 90) having no nilpotent elements due to (1.22'). Thus $K_i \otimes S$ is a division ring S_i with $x^{-1}S_i x = S_i$. Hence we can write

$$QT_0 = Q(A_1 \times K_8) = \sum^{\oplus} D_i ,$$

a direct sum of division rings with the corresponding idempotents central in QG.

Since σ_x maps QT_0 onto QT_0 and the corresponding idempotents are central it follows that

$$a\sigma_x(a) = 0 \Rightarrow a = 0. \quad \square$$

Corollary 1.24. If G is a nilpotent or FC-group then QG has no nilpotent elements if and only if (1.21) and (1.22) or (1.22') hold.

Proof: Theorem 1.20 and Proposition 1.8.

Remark: The condition (1.23) when specialized to group rings simply says:

R has no nilpotent elements $\Rightarrow$ $R(G/T)$ has no nilpotent elements,

which is essentially the famous Conjecture 1.3.

2. TORSION FREE $\mathcal{U}_1\mathbb{Z}G$

We shall prove that the only torsion units of $\mathbb{Z}G$ are ± 1 if and only if G is torsion free. We recall that $\mathcal{U}_1 KG$ denotes the group of 1-units of KG, namely

$$\mathcal{U}_1 KG = \left\{ \gamma = \sum \gamma_g g \in \mathcal{U}KG : \sum \gamma_g = 1 \right\} .$$

Theorem 2.1 (Sehgal [8]). Suppose that R is an integral domain of characteristic zero in which no rational prime is invertible. If U_1RG has an element $\gamma = \Sigma\gamma(g)g$ of order p^α, p a prime, then G has an element g_0 of order p^α with $\gamma(g_0) \neq 0$.

Proof: We have

$$1 = \gamma^{p^\alpha} \equiv \sum \gamma(g)^{p^\alpha} + \beta \pmod{pRG}$$

for some $\beta \in [RG,RG]$ by (I.1.4). Also by (I.1.5) we know that $\beta(1)$, the coefficient of identity in β, is zero. We have, therefore,

$$1 \equiv \sum_{g^{p^\alpha}=1} \gamma(g)^{p^\alpha} \pmod{pR}$$

$$\equiv \sum_{o(g)=p^\alpha} \gamma(g)^{p^\alpha} + \sum_{g^{p^{\alpha-1}}=1} \gamma(g)^{p^\alpha} \pmod{pR}$$

$$\equiv \sum_{o(g)=p^\alpha} \gamma(g)^{p^\alpha} + \left(\sum_{g^{p^{\alpha-1}}=1} \gamma(g)^{p^{\alpha-1}}\right)^p \pmod{pR}.$$

But since the order of γ is p^α we have by (II.1.2)

$$\gamma^{p^{\alpha-1}}(1) \equiv \sum_{g^{p^{\alpha-1}}=1} \gamma(g)^{p^{\alpha-1}} \equiv 0 \pmod{pR}.$$

Thus we conclude that

$$\sum_{o(g)=p^\alpha} \gamma(g) \not\equiv 0 \pmod{pR}.$$

Hence there exists an element $g_0 \in G$ of order p^α with $\gamma(g_0) \neq 0$. □

Corollary 2.2. Suppose that R is an integral domain of characteristic zero in which no rational prime is invertible. Suppose that there is an element $\gamma \in \mathcal{U}_1RG$ such that

$$o(\gamma) = p_1^{\alpha_1} p_2^{\alpha_2} \cdots p_k^{\alpha_k}$$

with p_i distinct primes. Then there exist elements $g_i \in G$ with $o(g_i) = p_i^{\alpha_i}$ for each i. Moreover, if G is torsion free then every torsion unit γ of RG is of the form $\gamma = \xi \in R$.

We have the

<u>Theorem 2.3</u> (Sehgal). $\mathcal{U}_1\mathbb{Z}G$ is torsion free if and only if G is torsion free.

The same argument as in (2.1), by using (II.1.4) instead of (II.1.2), gives us the following result which was proved for finite groups by Zassenhaus [2].

<u>Theorem 2.4</u>. Suppose that G is a polycyclic by finite group and R an integral domain of characteristic zero satisfying $\mathcal{U}R \cap \{o(g): g \in G\} = \{1\}$. If $\mathcal{U}_1KG$ has an element of order

$$n = p_1^{\alpha_1} p_2^{\alpha_2} \cdots p_k^{\alpha_k}$$

with p_i distinct primes then there exist elements $g_i \in G$ with $o(g_i) = p_i^{\alpha_i}$.

3. NILPOTENT UNIT GROUPS

In this section we shall give necessary and sufficient conditions for $\mathcal{U}KG$ to be nilpotent if K is a field or the ring of rational integers. This study was begun by Bateman and Coleman [1] who characterized group algebras KG, of finite groups G over a field K, with nilpotent unit groups. This was extended to arbitrar groups by Khripta ([1],[2]) and Fisher, Paramenter and Sehgal [1]. Polcino Milies [2] described finite groups G whose integral group rings have nilpotent unit groups. This work was completed by Sehgal and Zassenhaus [2].

Theorem 3.1. Let G be a group having an element h of order p in its center $\zeta(G)$. If K is a field of characteristic p, then

$$\mathcal{U}KG \text{ is nilpotent} \iff G \text{ is nilpotent with } G' \text{ a finite } p\text{-group.}$$

Lemma 3.2 (Khripta). With the same assumption as (3.1)

$$\mathcal{U}KG \text{ nilpotent} \Rightarrow \text{there exists an } m \text{ such that } G^{p^m} \subseteq \zeta(G).$$

Proof: Let us pick $x,y \quad G$ and write $\eta = \Sigma_1^p h^i$. Then $\eta^2 = 0$ and $1 + \eta x$ is a unit of KG with inverse $1 - \eta x$. Therefore,

$$\begin{aligned}(1+\eta x,y) &= (1+\eta x)^{-1}y^{-1}(1+\eta x)y \\ &= 1 + \eta(x^y - x), \quad x^y = y^{-1}xy; \\ (1+\eta x,y,y) &= 1 + \eta(x^{y^2} - 2x^y + x).\end{aligned}$$

By induction we get

$$\begin{aligned}(1+\eta x,y,\cdots,y) &= (1+\eta x,{}_n y) \\ &= 1 + \eta\left(x^{y^n} - \binom{n}{1}x^{y^{n-1}} + \binom{n}{2}x^{y^{n-2}} + \cdots + (-1)^n x\right).\end{aligned}$$

Choosing ℓ such that p^ℓ is larger than the class of nilpotence of $\mathcal{U}KG$ we conclude that

$$1 = (1+\eta x, {}_{p^\ell}y) = 1 + \eta(x^z - x), \quad z = y^{p^\ell}.$$

It follows that $\eta(x^z - x) = 0$ and thus there is an i such that $x^z = xh^i$. Consequently, $x^{z^p} = x$ and $(y^{p^{\ell+1}},x) = 1$. Since x and y are arbitrary and ℓ is independent of x and y we have proved the lemma. □

Lemma 3.3. Let K be a field and I a nilpotent ideal of KG. Then the natural projection $KG \to KG/I$ induces an epimorphism $\mathcal{U}KG \to \mathcal{U}(KG/I)$.

Proof: Clearly we have a homomorphism $\mathcal{U}KG \to \mathcal{U}(KG/I)$ induced by the natural projection $KG \to KG/I$. Suppose we have $\bar{\gamma} \in U(KG/I)$. Then there is a $\bar{\mu} \in U(KG/I)$ such that $\bar{\gamma}\bar{\mu} = \bar{1}$ which implies that $\gamma\mu - 1 = \delta \in I$. Since I is nilpotent $\delta^{k+1} = 0$ for some k. We conclude that

$$\gamma\mu\left(1-\delta+\delta^2+\cdots+(-1)^k\delta^k\right) = 1, \quad \gamma \in UKG. \quad \square$$

3.4. Proof of Theorem 3.1

(i) Necessity: Suppose UKG is nilpotent of class n. Then by Lemma 3.2 there exists an m such that $G^{p^m} \subseteq \zeta(G)$. It follows by Schur's theorem (I.4.3) that $(G')^{p^M} = 1$ for a fixed M.

To prove that G' is finite we use induction on the class of nilpotence of G. Thus we may assume that G'' is a finite p-group. In view of (3.3), factoring by G'' we may further assume that G' is an abelian p-group. Suppose that G' is infinite. Then there exists an r such that $\gamma_r(G)$ is infinite but $\gamma_{r+1}(G)$ is finite. Again factoring by $\gamma_{r+1}(G)$ we can conclude that G' contains an infinite central group A. Thus we have:

$$G' \text{ Abelian}, \quad (G')^{p^M} = 1, \quad G' \supseteq A, \quad A \subseteq \zeta(G), \quad |A| = \infty.$$

We claim

(3.5) Given any $0 \neq \delta \in KG$ and a natural number t there exist central elements $\mu_i \in KG$ with $\mu_i^2 = 0$ but $\mu_1\mu_2\cdots\mu_t\delta \neq 0$.

To see this write

$$\delta = g_1\alpha_1 + \cdots + g_s\alpha_s, \quad \alpha_i \in KA, \quad \alpha_i \neq 0$$

where g_i's belong to different cosets of A. Let B be the finite group generated by the supports of α_i, $1 \le i \le s$. Then $B \subsetneq A$. Since A is infinite Abelian, of bounded exponent it can be written as a product of cyclic groups (see Fuchs [1]). Hence we can write

$$A = B_1 \times A_1 \times A_2 \times \cdots \times A_t \times C_1$$

where B_1 is finite containing B, A_i are finite and C_1 is a suitable subgroup. Setting $\mu_i = \Sigma_{a\in A_i} a$ we conclude that $\mu_i^2 = 0$ and

$$\mu_1\mu_2 \cdots \mu_t\delta \neq 0.$$

Now observe that for any $\gamma_i \in KG$, $(1+\mu_i\gamma_i) \in \mathcal{U}KG$ and

$$(1+\mu_1\gamma_1,1+\mu_2\gamma_2) = 1 + \mu_1\mu_2[\gamma_1,\gamma_2].$$

By induction,

$$(1+\mu_1\gamma_1,1+\mu_2\gamma_2,\cdots,1+\mu_t\gamma_t) = 1 + \mu_1\mu_2\cdots\mu_t[\gamma_1,\cdots,\gamma_t].$$

Since G' is infinite it follows by (V.4.4) that KG is not Lie nilpotent. Therefore for any t there exist γ_i such that $[\gamma_1,\cdots,\gamma_t] \neq 0$. It follows by (3.5) that $\mathcal{U}KG$ is not nilpotent. This is a contradiction proving that G' is finite.

(ii) "Sufficiency" is proved in (V.6.16) by observing

(a) $\gamma_n(\mathcal{U}KG) - 1 \subseteq (KG)^{(n)}$

and

(b) G nilpotent, G' finite p-group $\Rightarrow$ $(KG)^{(n)} = 0$ for some n. □

If a nilpotent group has a p-element then it certainly has one in its center. Thus it remains to characterize group algebras KG with nilpotent unit groups when G has no p-element and K has characteristic $p > 0$. This is done in the next theorem which was proved for $p \neq 2,3$ by Fisher, Parmenter and Sehgal [1] and by Khripta in her thesis as announced by Zalesskii and Mikhalev [1]. Passman has pointed out that the restriction $p \neq 2,3$ is not necessary. Before we state the main result let us recall that we denote by $T(G)$ the set of all torsion elements of G. Also we say that G is n-Engel if it satisfies $(x,_n y) = 1$ for all $x,y \in G$. So, certainly, a nilpotent group of class n is n-Engel.

<u>Theorem 3.6.</u> Suppose KG is a group algebra over a field K of

characteristic $p \geq 0$. Suppose G has no element of order p (if $p > 0$). Then $\mathcal{U}KG$ is nilpotent if and only if G is nilpotent and one of the following holds.

(a) $T(G)$ is a central subgroup,

(b) $|K| = 2^\beta - 1 = p$, a Mersenne prime; $T(G)$ is an Abelian group of exponent (p^2-1) and for $x \in G$, $t \in T(G)$, $x^{-1}tx = t$ or t^p.

We shall need several lemmas some of which will be used later as well.

Lemma 3.7. Suppose that $\mathcal{U}RG$ is nilpotent where R is an integral domain of characteristic $p \geq 0$ and G has no p-element (if $p > 0$) Then every subgroup of $T(G)$ is normal in G. Moreover, if R is a field then $T(G)$ is an Abelian subgroup.

Proof: Let t be a torsion element, of order m, in G. For any $x \in G$, consider

$$\mu = (1-t)x(1+t+\cdots+t^{m-1}).$$

Clearly, $\mu^2 = 0$, $(1-\mu)(1+\mu) = 1$ and $1 - \mu \in \mathcal{U}RG$. We have

$$\left((1-\mu),t^{-1}\right) = (1+\mu)t(1-\mu)t^{-1} = (t+\mu)(t^{-1}-\mu) = 1 - (t-1)\mu.$$

By induction it follows that for all $k \geq 1$

$$(1-\mu,t^{-1},\cdots,t^{-1}) = (1-\mu,_k t^{-1}) = 1 - (t-1)^k\mu. \tag{3.8}$$

Let us make another observation.

$$(t-1)^2\alpha = 0, \quad \alpha \in RG \Rightarrow (t-1)\alpha = 0 \tag{3.9}$$

This follows because

$$\begin{aligned}(t-1)^2\alpha = 0 \Rightarrow (t-1)\alpha &= \left(\sum_1^m t^i\right)\beta \quad \text{for some } \beta \in RG \\ \Rightarrow 0 &= \left(\sum t^i\right)(t-1)\alpha = m\left(\sum t^i\right)\beta \\ \Rightarrow 0 &= \left(\sum t^i\right)\beta = (t-1)\alpha \ .\end{aligned}$$

Since $\mathcal{U}RG$ is nilpotent, there exists an n, by (3.8), such that

$$(t-1)^n\mu = 0 \overset{(3.9)}{\Longrightarrow} (t-1)\mu = 0 \overset{(3.9)}{\Longrightarrow} \mu = 0.$$

Thus $(1-t)x(\Sigma t^i) = 0$ which implies that there exists a j such that $tx = xt^j$ and consequently $\langle t\rangle \triangleleft G$. Hence $T(G)$ is an Abelian or a Hamiltonian group.

It remains to prove:

> If F is a prime field with $|F| \neq 2$ then $\mathcal{U}FK_8$ is not nilpotent.

Since $\mathcal{U}QK_8$ is not nilpotent as seen in (1.14) we let $F = \mathbb{Z}/p\mathbb{Z}$, $p \neq 2$. Then by Wedderburn's structure theorem,

$$FK_8 \simeq \sum^{\oplus} (D_i)_{n_i} ,$$

a direct sum of full matrix rings $(D_i)_{n_i}$ over division rings D_i. Since K_8 is not commutative and each D_i, being finite, is commutative at least one n_i say $n_0 > 1$. It follows that $\mathcal{U}FK_8$ is not nilpotent as $\mathcal{U}\left((D_0)_{n_0}\right)$ is not. □

Corollary 3.10. If $\mathcal{U}\mathbb{Z}G$ is nilpotent then every idempotent of QA, where A is an Abelian torsion subgroup of G, is central in QG.

Proof: By (1.16) every idempotent e of QA is a rational linear combination of the idempotents of the type

$$\varepsilon_C = \frac{1}{|C|} \sum_{g \in C} g$$

where C runs over subgroups of A. Since every C is normal in G by the last lemma, ε_C and thus e is central in QG. □

The next lemma is well known (see, for example, Jacobson [2] or Passman [6]).

Lemma 3.11. Let R be a ring which contains a set of elements e_{ij}, $1 \leq i,j \leq n$, satisfying

$$1 = \sum_1^n e_{ii}, \quad e_{ij}e_{jk} = e_{ik}, \quad e_{ij}e_{\ell k} = 0 \quad \text{for} \quad \ell \neq j.$$

Then $R \simeq M_n(S)$, the full ring of $n \times n$ matrices over S, where S is the centralizer of e_{ij}, $1 \le i,j \le n$. Moreover, $S \simeq e_{11}Re_{11}$.

Proof: Given $a \in R$, set $b_{ij} = b_{ij}(a) = \Sigma_k e_{ki} a e_{jk}$. Then

$$b_{ij}e_{\ell m} = e_{\ell i} a e_{jm} = e_{\ell m}b_{ij},$$

which implies that $b_{ij} \in S$. We define a map

$$\sigma: R \to M_n(S)$$

by $\sigma(a) = (b_{ij})$. Suppose that $\sigma(a') = (b'_{ij})$. Then clearly $\sigma(a+a') = \sigma(a) + \sigma(a')$. Moreover,

$$\sum_\lambda b_{i\lambda}b'_{\lambda j} = \sum_\lambda \left(\sum_k e_{ki} a e_{\lambda k}\right)\left(\sum_\ell e_{\ell\lambda} a' e_{j\ell}\right)$$

$$= \sum_{k,\lambda} e_{ki} a e_{\lambda\lambda} a' e_{jk}$$

$$= \sum_k e_{ki} a a' e_{jk}$$

which implies that

$$\sigma(aa') = (b_{ij})(b'_{ij}) = \sigma(a)\sigma(a').$$

Thus σ is a ring homomorphism. Suppose that $\sigma(a) = 0$ then

$$0 = \sum_{i,j} b_{ij}e_{ij} = \sum_{k,i,j} e_{ki} a e_{jk}e_{ij} = \sum_{i,j} e_{ii} a e_{jj} = a.$$

Thus σ is 1-1. Now given $(c_{ij}) \in M_n(S)$ it is easy to check that

$$\sigma\left(\sum c_{ij}e_{ij}\right) = (c_{ij}).$$

We have proved that $R \simeq M_n(S)$. To prove that S is isomorphic to $e_{11}Re_{11}$ we have only to observe that $\sigma(e_{ij})$ is indeed the matrix which has 1 in the (i,j) place and zeros elsewhere and that the centralizer of the matrices, $\sigma(e_{ij})$, $1 \leq i,j \leq n$, in $M_n(S)$ is S which is isomorphic to $\sigma(e_{11})M_n(S)\sigma(e_{11})$. □

The last result has the following useful consequence in the present context (see also Passman [6]).

Lemma 3.12. Suppose that K is a field of characteristic $p \geq 0$ and that A is a finite normal Abelian subgroup of G. Suppose that A has no element of order p (if $p > 0$) and that there exists an idempotent of KA which is not central in KG. Then $\mathcal{U}KG$ contains an isomorphic copy of the full linear group, $\mathcal{U}M_m(F)$, over a field $F \supseteq K$ for some $m > 1$.

Proof: We may suppose without loss of generality that $G = \langle A,x \rangle$. We have by hypothesis,

$$KA = F_1 \oplus \cdots \oplus F_s ,$$

a direct sum of fields with corresponding primitive orthogonal idempotents $\{e_i\}$. Also by hypothesis at least one of the e_i's, say e_1 is not central, $e_1^x = x^{-1}e_1x \neq e_1$. Because of the normality of A, we may arrange the subscripts so that

$$e_1^x = e_2, \quad e_2^x = e_3, \cdots, e_m^x = e_1, \quad m > 1.$$

Then

$$e = e_1 + e_2 + \cdots + e_m$$

is a central idempotent of KG. It is, therefore, enough for us to prove that $\mathcal{U}(eKG) \supseteq \mathcal{U}M_m(F)$ for some F. Setting

$$e_{ij} = x^{-(i-1)}e_1\, x^{(j-1)} \quad \text{for} \quad 1 \leq i,j \leq m$$

we see that

$$e_{ii} = e_i, \quad \sum_1^m e_{ii} = e, \quad e_{ij}e_{jk} = e_{ik}.$$

Also, for $\ell \neq j$

$$e_{ij}e_{\ell k} = x^{-(i-1)}e_1x^{(j-1)}x^{-(\ell-1)}e_1x^{(k-1)} = x^{(j-i)}e_je_\ell x^{(k-\ell)} = 0.$$

It follows by the last lemma that

$$eKG \simeq M_m(S), \quad S \simeq e_1(eKG)e_1 = e_1(KG)e_1.$$

But $e_1(KG)e_1 \supseteq e_1(KA)e_1 = F_1$ and both have the same identity. Hence $\mathcal{U}(eKG)$ contains an isomorphic copy of $\mathcal{U}(M_m(F_1))$. □

Corollary 3.13. Suppose that K is a field of characteristic $p \geq 0$ and G has no element of order p (if $p > 0$). Then

$\mathcal{U}KG$ nilpotent $\Rightarrow$ every idempotent of KG is central.

Proof: First of all, by (3.7) we conclude that $T = T(G)$, the torsion subgroup of G, is Abelian with every subgroup normal in G. Then it follows by (3.12) that every idempotent of KT is central in KG. It remains to prove that every idempotent e of KG actually belongs to KT. Replacing G by the group generated by the support of e we may assume that G is finitely generated and T is finite. Then

$$KT = F_1 \oplus \cdots \oplus F_s,$$

a direct sum of fields. Therefore,

$$KG = (KT)(G/T,\rho,\sigma)$$

a crossed product of G/T over KT with natural ρ and σ. Due to the centrality of idempotents of KT we have

$$KG = \sum^{\oplus} F_i\ (G/T,\rho,\sigma)$$

$$= \sum^{\oplus} F_i(G/T,\rho_i,\sigma_i)$$

with ρ_i, σ_i the natural projections of ρ and σ. It follows by (1.7) that $e \in \Sigma^{\oplus} F_i = KT$. □

Lemma 3.14 (Birkhoff and Vandiver). Let n and j be natural numbers > 1. Then every prime divisor of $n^j - 1$ divides $n-1$ if and only if $j = 2$ and $n = 2^m - 1$ for some m.

Proof: If $n = 2^m - 1$ then $n^2 - 1 = 2^m(n-1)$, $n - 1 = 2(2^{m-1}-1)$ and the sufficiency of the condition is clear. Conversely, let us assume that every prime divisor of $n^j - 1$ divides $n - 1$. First suppose that j is an odd prime. From the equation

$$(n^j-1)/(n-1) = \big((1+n-1)^j-1\big)/(n-1) = j + \sum_{i=2}^{j}\binom{j}{i}(n-1)^{i-1} \qquad (*)$$

we deduce that the only prime divisor of $(n^j-1)/(n-1)$ is j and that

$$(n^j-1)/(n-1) = j^s, \quad s > 1.$$

It follows from (*) that j^2 divides j as every other term on the right is a multiple of j^2. This is a contradiction proving that j can't be an odd prime.

Now from the observation

$$n^{ab} - 1 = (n^a)^b - 1$$

and the fact that $n^a - 1$ is a multiple of $n - 1$ we may conclude that $j = 2^\ell$ for some ℓ. Therefore, $n^j - 1$ is a multiple of $n^2 - 1$. Consequently any divisor of $n + 1$ divides $n - 1$ and hence is 2. Thus $n + 1 = 2^m$. It remains to prove that $j = 2$. Suppose to the contrary that $j > 2$. Then since $n^j - 1$ is a multiple of $n^2 + 1$, every prime divisor of $n^2 + 1$ divides $n - 1$ and therefore $n^2 - 1$. It follows that

$$n^2 + 1 = 2^t, \quad n + 1 = 2^m, \quad m > 1$$

and

$$2^t = (2^m-1)^2 + 1 = 2^{2m} - 2^{m+1} + 2.$$

This is a contradiction proving the lemma. □

Lemma 3.15. If p is a rational prime such that $2^j - 1 = p^n$ for some natural numbers j and n then $n = 1$.

Proof: Suppose that n is even; then

$$2^j - 1 = p^n \equiv 1 \pmod 8$$

which is a contradiction and thus n is odd. Now,

$$2^j = p^n + 1 = (p+1)(p^{n-1} - p^{n-2} + \cdots + 1)$$

implies that $n = 1$ as otherwise the second factor, being a sum of an odd number of odd integers, is an odd number dividing 2^j. □

Let F be a field contained in KG. Suppose that $x \in G$ has infinite order, $\langle x \rangle$ is linearly independent over F and that x induces an automorphism $\alpha = \alpha_x$ of F by $f \to xfx^{-1} = f^\alpha$. Then we have an isomorphic copy of the skew group ring $F_\alpha\langle x \rangle$ contained in KG. Thus

$$F_\alpha\langle x \rangle = \left\{ \sum f_i x^i \mid f_i \in F \right\}, \quad xf = f^\alpha x.$$

We shall keep this notation fixed and also denote by $\dot{F}$ the multiplicative group of F.

Lemma 3.16. For all $f \in \dot{F}$ we have

$$(f,_m x^{-1}) = (f, x^{-1}, \cdots, x^{-1}) = f^{(\alpha-1)^m} .$$

Proof: We use induction on m. Clearly

$$(f, x^{-1}) = f^{-1} x f x^{-1} = f^{-1} f^\alpha = f^{(\alpha-1)}.$$

Suppose that the formula holds for m. Then

$$\begin{aligned}((f,_m x^{-1}), x^{-1}) &= (f^{(\alpha-1)^m}, x^{-1}) \\ &= f^{-(\alpha-1)^m} x\, f^{(\alpha-1)^m} x^{-1} \\ &= f^{-(\alpha-1)^m} f^{\alpha(\alpha-1)^m} = f^{(\alpha-1)^{m+1}} .\end{aligned}$$

The lemma is proved. □

Lemma 3.17. Suppose that in $F_\alpha\langle x\rangle$ we have

$$(f,{}_m x^{-1}) = (f,x^{-1},x^{-1},\cdots,x^{-1}) = 1$$

for all $f \in \dot{F}$ and $\alpha \neq I$ is of finite order. Then $|F| = p^2$ where p is a Mersenne prime and $f^\alpha = f^p$ for all $f \in F$.

Proof: First suppose that F is infinite. We know by the last lemma that

$$f^{(\alpha-I)^m} = 1, \quad f^{(\alpha^s-I)} = 1$$

for all $f \in \dot{F}$, where s is the order of α. There exist polynomials $u(x)$, $v(x) \in \mathbb{Z}[x]$ and $a \in \mathbb{Z}$, $a \neq 0$ such that

$$(x-1)^m u(x) + (x^s-1)v(x) = a(x-1)$$

which implies that

$$(\alpha-I)^m u(\alpha) + (\alpha^s-I)v(\alpha) = a(\alpha-I).$$

It follows that for all $f \in \dot{F}$ we get

$$1 = f^{a(\alpha-I)}, \quad (f^a)^\alpha = f^a.$$

Clearly $|F^a|$ is infinite. Let K be the fixed field of α. Then

$$K \supseteq F^a, \quad F = K(\beta) \text{ for some } \beta.$$

Also, $x^a = b \in F^a$ is an equation satisfied by β and

$$\beta^a = (\beta^a)^\alpha = (\beta^\alpha)^a$$

implies that $\beta^\alpha = \beta\xi$ where ξ is an a-th root of unity. Similarly for any $r \in F^a$ we have $(\beta+r)^\alpha = (\beta+r)\eta$ where η is an a-th root of unity. Since $|F^a|$ is infinite, there exists an infinite number of $r \in F^a$ and a fixed η such that $(\beta+r)^\alpha = (\beta+r)\eta$. But $(\beta+r)^\alpha = \beta^\alpha+r = \beta\xi + r$ which implies $r(\eta-1) = \beta(\xi-\eta)$. Since this equation holds for an infinity of values of r we can conclude that $\eta - 1 = 0 = \xi - \eta$. Hence $\beta^\alpha = \beta$ and $\alpha = I$ which is a contradiction. Hence F is finite.

Suppose that $|F| = p^a$. Since $f^\alpha = f^{p^j}$ for some $j < a$ we have that

$$(f,_m x^{-1}) = f^{(\alpha-1)^m} = f^{(p^j-1)^m} = 1$$

for all $f \in \dot{F}$. Therefore, (p^a-1) divides $(p^j-1)^m$. Hence,

$$\text{any prime divisor of } (p^a-1) \text{ divides } (p^j-1). \qquad (*)$$

We claim that (*) implies $a = 2$. Let j be the smallest natural number such that (*) holds for a fixed a. Then writing $a = jq + r$,.

$$p^a - 1 = p^{jq+r} - 1 = p^r(p^{jq}-1) + (p^r-1).$$

It follows that any prime divisor of $p^a - 1$ is a divisor of $p^r - 1$. We may thus assume that $a = jq$. Thus any prime divisor of $(p^j)^q - 1$ is a divisor of $p^j - 1$. From (3.14) and (3.15) we deduce

$$q = 2, \quad p^j = 2^\gamma - 1, \quad j = 1.$$

Hence $a = jq = 2$. We have proved that $|F| = p^2$, $p = 2^\gamma - 1$ and hence $f^\alpha = f^p$. □

3.18. <u>Proof of Theorem 3.6</u>

(i) Necessity. Let us assume that $\mathcal{U}KG$ is nilpotent. Then, certainly, G is nilpotent. We also know by (3.7) that $T = T(G)$ is an Abelian group with every subgroup normal in G. Suppose that T is not central. Then we wish to prove that (b) holds. We may, therefore, assume that

$$G = \langle T,x\rangle, \quad |T| < \infty, \quad o(x) = \infty, \quad T \text{ not central.}$$

By Wedderburn's structure theorem

$$KT = F_1 \oplus \cdots \oplus F_k, \qquad (3.19)$$

a direct sum of fields. Thus

$$KG = (KT)_\alpha\langle x\rangle = \left(\sum^{\oplus} F_i\right)_\alpha \langle x\rangle = \sum^{\oplus}(F_i)_\alpha\langle x\rangle.$$

The last equality holds because all idempotents are central by (3.13) and $F_i^x = F_i$. It follows by (3.17) that K is finite with $|K| = p$ or p^2, $p = 2^\beta - 1$. If $|K| = p^2$ then

$$|F_i| = |K| \Rightarrow F_i = eKT = eK, \quad e^2 = e.$$

This implies due to the centrality of e that x commutes with F_i. But since x is not central, $|K|$ must be p. It remains to prove that $T^{(p^2-1)} = 1$ and

(3.20) $\quad xt \neq tx \Rightarrow x^{-1}tx = t^p.$

Suppose that $T^{(p^2-1)} \neq 1$. Choose $g,h \in T$ with

$$x^{-1}hx = h^x \neq h \quad \text{and} \quad g^{(p^2-1)} \neq 1.$$

Then

$$y = (1-g^{p^2-1})(1-h^x h^{-1}) \neq 0$$

since the coefficient of identity in this product is 1 or 2 and $p \neq 2$. Therefore, there exists an F_i and a homomorphism $\lambda\colon KT \to F_i$ with $\lambda(y) \neq 0$. Thus $\lambda(g)^{p^2-1} \neq 1$ and $|F_i| > p^2$. But,

$$1 \neq \lambda(h^x h^{-1}) = \lambda(h^x)\lambda(h^{-1}) = \lambda(h)^x \lambda(h^{-1})$$

implies that F_i not central in contradiction with Lemma 3.17. We have proved that $T^{(p^2-1)} = 1$ and it remains to prove (3.20).

We observe that, since $p = 2^\beta-1$, $p^2-1 = 2^\beta(p-1)$, if for some $\xi \in F_i$, $\xi^{p^2-1} = 1$, $\xi^{p-1} \neq 1$ then $o(\xi)$ is a multiple of 4. Suppose that contrary to (3.20) we have $t^x \neq t$, t^p. Then in (3.19), after rearranging the F_i's, we have

$$t = (\xi,\eta,\cdots), \quad t^x = (\xi^p,\eta,\cdots), \quad \xi^p \neq \xi, \quad \eta^p \neq \eta.$$

On the other hand,

$$t^x = t^i = (\xi^i,\eta^i,\cdots)$$

which implies that $p-i = 0 \pmod 4$ and $i-1 \equiv 0 \pmod 4$. Thus $p-1 \equiv 0 \pmod 4$ which is a contradiction proving (3.20).

(ii) Sufficiency. We now assume that G is nilpotent of class c. We wish to prove that $\mathcal{U}KG$ is nilpotent of class at most $(c+1)$ or $(c+\beta+1)$ according as (a) or (b) is satisfied. In any case, we may assume that G is finitely generated and thus T is finite. Therefore $KT = \Sigma^{\oplus} F_i$, a finite direct sum of fields. We assert that every idempotent of KT is central in KG. This is trivial if (a) holds. So let us assume that (b) holds. Let $e = e^2 = \Sigma e_g g \in KT$. Then

$$e = e^p = \sum e_g g^p \Rightarrow e_g = e_{g^p} .$$

Therefore, since $g^x = g$ or g^p we have $e^x = e$. Now since

$$KG = (KT)(G/T,\rho,\sigma) = \left(\sum^{\oplus} F_i\right)(G/T,\rho,\sigma) = \sum^{\oplus} F_i(G/T,\rho_i,\sigma_i)$$

it follows by (1.7) that we have the direct product decomposition

$$\mathcal{U}KG = \Pi \dot{F}_i \cdot G/T.$$

It suffices to prove that $\dot{F}_i \cdot G/T$ is nilpotent of class at most $(c+1)$ or $(c+\beta+1)$. This is clear if σ is trivial i.e. if $\dot{F}_i$ and G/T commute. We may, therefore, suppose that we have $|F_i| = p^2$, $p = 2^\beta - 1$. We wish to prove that $\dot{F}_i \cdot G/T$ is nilpotent of class $\le (c+\beta+1)$. It is easy to check by direct computation that $\dot{F}_i \subseteq \zeta_{\beta+1}$, the $(\beta+1)$-th term of the upper central series of $(\dot{F}_i \cdot G/T)$. Since G/T is nilpotent of class c it follows that $\dot{F}_i \cdot G/T$ is nilpotent of class $\le (c+\beta+1)$. □

It has been proved by Fisher, Parmenter and Sehgal [1] that if K is a field of characteristic $p \ge 0$ and G has no p-element (if $p > 0$) then $\mathcal{U}KG$ is solvable, n-Engel for some n if and only if $\mathcal{U}KG$ is nilpotent.

Next, we shall characterize groups G such that $\mathcal{U}\mathbb{Z}G$ is nilpotent. We already know by Lemma 3.7 that if $\mathcal{U}\mathbb{Z}G$ is nilpotent then every torsion subgroup of G is normal. Thus, either the torsion subgroup T is Abelian or $T = A \times E \times K_8$ where A is Abelian with every element of odd order, $E^2 = 1$ and K_8 is the quaternion group. Moreover, if T_1 is any Abelian torsion subgroup of G then every idempotent of QT_1 is central in QG ((3.10)). The following lemma is the key result in our investigation of the problem at hand.

Lemma 3.21. Suppose that $\mathcal{U}\mathbb{Z}G$ is nilpotent and $t, t_1, t_2 \in T$, $x \in G$. Then

(i) $x^{-1}tx \neq t \Rightarrow x^{-1}tx = t^{-1}$;

(ii) $o(t)$ odd $\Rightarrow xt = tx$;

(iii) $o(t_1)$ odd > 1, $o(t_2)$ even $\Rightarrow T$ central in G.

Proof: (i) Suppose t has order $m > 1$ then by (II.2.6)

$$Q\langle t\rangle = \sum_{d|m}^{\oplus} Q(\xi_d),$$

where ξ_d is a primitive d-th root of unity. This induces an epimorphism

$$\mathbb{Z}\langle t\rangle \to R = \sum_{d|m}^{\oplus} \mathbb{Z}[\xi_d] \to \mathbb{Z}[\xi_m],$$

where

$$t \to \sum \xi_d \to \xi_m.$$

Also, we know by (II.2.9) that

$$(\mathcal{U}R\colon \mathcal{U}\mathbb{Z}G) = s < \infty, \quad (\mathcal{U}R)^s \leq \mathcal{U}\mathbb{Z}G.$$

It follows that $V = (\mathcal{U}\mathbb{Z}[\xi_m])^{2ms}$ is a torsion free subgroup of $\mathcal{U}\mathbb{Z}G$ which is invariant under conjugation by x. This is because every idempotent of $Q\langle t\rangle$ is central in QG which, of course, implies

$$x^{-1}Q(\xi_d)x = Q(\xi_d), \quad x^{-1}\mathbb{Z}[\xi_d]x = \mathbb{Z}[\xi_d].$$

Certainly, if $m = 2,3,4$ or 6 then $x^{-1}tx = t$ or t^{-1} due to the normality of $\langle t\rangle$. Otherwise, $\mathcal{U}\mathbb{Z}[\xi_m]$ is not torsion and V is a nontrivial finitely generated torsion free Abelian group with $\langle V,x\rangle$ nilpotent. We may choose a basis of V and represent x by an integral matrix X. The computation in (3.16) shows that $(X-I)$ is a nilpotent matrix and therefore has all its characteristics values zero. Thus X is a matrix of finite multiplicative order with all its characteristic values equal to 1. Therefore $X = I$ and x fixes element wise the group V which has the same rank as $\mathcal{U}\mathbb{Z}[\xi_m]$. It follows by (II.2.10) that either $Q(V) = Q(\xi_m)$ and $x^{-1}\xi_m x = \xi_m$ or $Q(V)$ is the fixed field of the automorphism $\xi_m \to \xi_m^{-1}$ and $x^{-1}\xi_m x = \xi_m^{-1}$. Further, because $t = \Sigma\xi_d$ implies $x^{-1}tx = t^i = \Sigma\xi_d^i$, it follows that

$$x^{-1}tx = t \text{ or } t^{-1} \text{ according as } x^{-1}\xi_m x = \xi_m \text{ or } \xi_m^{-1}.$$

(ii) Suppose that m is odd and $x^{-1}tx = t^{-1}$ then

$$(t,x,x,\cdots,x) = t^{\pm 2^i} \neq i.$$

It follows that $xt = tx$.

(iii) Suppose we have $x^{-1}t_1x = t_1$, $x^{-1}t_2x = t_2^{-1}$. Then $x^{-1}t_1t_2x = t_1t_2^{-1}$ but we should have by (i) $x^{-1}t_1t_2x = t_1t_2$ or $t_1^{-1}t_2^{-1}$. This implies that $t_2^2 = 1$ and thus $x^{-1}t_2x = t_2$. □

<u>Lemma 3.22</u>. Let G be an extension of a finite group T by a torsion free nilpotent group and K a field of characteristic 0 such that

$$KT = D_1 \oplus \cdots \oplus D_n,$$

a direct sum of division rings, with every idempotent of KT central in KG. Then

(i) every $u \in \mathcal{U}KG$ can be written as $u = \Sigma f_i g_i$, $f_i \in D_i$, $g_i \in G$;

(ii) $\mathcal{U}\mathbb{Z}G = (\mathcal{U}\mathbb{Z}T)\cdot G$.

<u>Proof</u>: Write $1 = \Sigma_1^n e_i$, $e_i \in D_i$, a sum of primitive orthogonal idempotents of KT and pick a transversal S of T in G such

that $s \in S \Rightarrow s^{-1} \in S$. For $u \in \mathcal{U}KG$ we have

$$u = \sum_j \alpha_j g_j, \quad \alpha_j \in KT, \quad g_j \in S$$

$$e_i u = \sum_j (e_i \alpha_j) g_j, \quad e_i \alpha_j \in D_i, \quad e_i u e_i u^{-1} = e_i.$$

Since G/T can be ordered, $e_i u \in D_i(G/T,\rho,\sigma)$ has just one term, i.e. $e_i u = e_i \alpha_k g_k$ (see (1.6)). Thus we have

$$u = \sum_i e_i u = \sum_i f_i g_i, \quad f_i \in \dot{D}_i, \quad g_i \in S.$$

Let us now suppose that $u \in \mathcal{U}\mathbb{Z}G$. Then we may write

$$u = \sum_1^n f_i g_i, \quad u^{-1} = \sum_1^n h_i \ell_i, \quad f_i, \ell_i \in \dot{D}_i, \quad g_i, h_j \in S.$$

Thus due to the fact that $x^{-1} D_i x = D_i$ for all $x \in G$ and all i

$$1 = uu^{-1} = \sum_1^n f_i (g_i h_i) \ell_i = \sum_1^n f_i (g_i h_i) \ell_i (g_i h_i)^{-1} (g_i h_i).$$

It follows that $g_i h_i = 1$ for all i. Thus collecting equal g_i's we can write

$$u = \sum \alpha(g) g, \quad u^{-1} = \sum h \beta(h)$$

with $\alpha(g), \beta(h) \in \mathbb{Z}T$, $g,h \in S$ satisfying for $i \neq j$ and $g^{-1} \neq h$

$$\alpha(g)\beta(h) = 0, \quad \alpha(g_i)\alpha(g_j) = 0 = \beta(h_i)\beta(h_j).$$

Also, we have

$$1 = \sum \alpha(g)\beta(g^{-1}).$$

Multiplying by a nonzero $\alpha(g_0)$ we obtain

$$\alpha(g_0) = \alpha(g_0)\beta(g_0^{-1})\alpha(g_0), \; e = \alpha(g_0)\beta(g_0^{-1}) = e^2 \neq 0.$$

Since $e \in \mathbb{Z}T$ it follows that $e = \alpha(g_0)\beta(g_0^{-1}) = 1 = \beta(g_0^{-1})\alpha(g_0)$.

Therefore, for $h \neq g_0^{-1}$ the equation $\alpha(g_0)\beta(h) = 0$ gives us $\beta(h) = 0$. We conclude that $u^{-1} = g_0^{-1}\beta(g_0^{-1})$ and hence

$$u = \alpha(g_0)g_0 \in (U\mathbb{Z}T) \cdot G. \quad \Box$$

The next theorem was proved for finite groups by Polcino Milies [2] and for arbitrary groups by Sehgal and Zassenhaus [2].

Theorem 3.23. $U\mathbb{Z}G$ is nilpotent if and only if G is nilpotent and the torsion subgroup T of G satisfies one of the following:

(i) T is central in G,

(ii) T is an Abelian 2-group and for $x \in G$, $t \in T$

$$x^{-1}tx = t^{\delta(x)}, \quad \delta(x) = \pm 1,$$

(iii) $T = E \times K_8$ where $E^2 = 1$ and K_8 is the quaternion group of order 8. Moreover, E is central in G and conjugation by $x \in G$ induces on K_8 one of the four inner automorphisms.

Proof: Suppose that $U\mathbb{Z}G$ is nilpotent. Then T is Abelian or Hamiltonian. If T is Abelian, not central, suppose

$$x^{-1}t_1x = t_1, \quad x^{-1}t_2x = t_2^{-1} \quad \text{for some } t_1, t_2 \in T.$$

Then $x^{-1}t_1t_2x = t_1t_2^{-1}$ which should be t_1t_2 or $t_1^{-1}t_2^{-1}$ due to (3.21). Hence either $t_1^2 = 1$ and $x^{-1}t_ix = t_i^{-1}$, $i = 1,2$ or $t_2^2 = 1$ and $x^{-1}t_ix = t_i$. In any case we have

$$x^{-1}tx = t^{\delta(x)} \quad \text{for all } t \in T$$

Moreover, T is a 2-group by (3.21). Thus we have the possibility (ii).

Now let us suppose that T is Hamiltonian. Then it follows by (3.21) that $T = E \times K_8$ where $E^2 = 1$. Every $x \in G$ maps every subgroup of K_8 onto itself. It follows easily that conjugation by x is, in fact, one of the four inner automorphisms of K_8.

It remains to prove that under the hypothesis of the theorem

$\mathcal{U}\mathbb{Z}G$ is nilpotent. In any case, it follows by (1.24) that QG has no nilpotent elements and thus by (1.9) every idempotent of QG is central. We conclude by (3.22) that

$$\mathcal{U}\mathbb{Z}G = (\mathcal{U}\mathbb{Z}T)\cdot G.$$

We consider the three cases separately.

(a) Assume (i). Then $\mathcal{U}\mathbb{Z}T$ is in the center of $\mathcal{U}\mathbb{Z}G$ and G is nilpotent. Hence $\mathcal{U}\mathbb{Z}G$ is nilpotent.

(b) Assume (ii). We claim that $\mathcal{U}\mathbb{Z}T$ is contained in ζ_m, the m-th term of the upper central series of $\mathcal{U} = \mathcal{U}\mathbb{Z}G$ where m is the nilpotency class of G. Let us first notice that if we have

$$u \in \mathcal{U}\mathbb{Z}T, \quad v = \tau x \in \mathcal{U}, \quad \tau \in \mathcal{U}\mathbb{Z}T, \quad x \in G$$

then

$$(u,v) = (u,\tau x) = (u,x) = u^{-1}u^{x} = \gamma \quad \text{where} \quad \gamma = 1 \quad \text{or} \quad u^{-1}u^{*}.$$

Here, for $u = \Sigma a_g g$ we have written $u^{*} = \Sigma a_g g^{-1}$. Since $\gamma^{*} = \gamma^{-1}$ it follows that $\gamma = \pm t$, $t \in T$. We have proved

$$(\mathcal{U}\mathbb{Z}T,\mathcal{U}) \subseteq \pm T$$

which implies that

$$(\mathcal{U}\mathbb{Z}T,\mathcal{U},\mathcal{U}) \subseteq \pm(T,\mathcal{U}) = \pm(T,G).$$

Repeating this we obtain

$$(\mathcal{U}\mathbb{Z}T,\mathcal{U},\mathcal{U},\cdots,\mathcal{U}) \subseteq \pm(T,G,G,\cdots,G) = 1.$$

It follows that $\mathcal{U}\mathbb{Z}T \subseteq \zeta_m$ and $\mathcal{U}\mathbb{Z}G$ is nilpotent.

(c) Assume (iii). In this case we know (II.2.5) that $\mathcal{U}\mathbb{Z}T = \pm T$ and hence $\mathcal{U}\mathbb{Z}G = \pm G$ is nilpotent. □

The last theorem has been used by Polcino Milies [3] to characterize group rings Z_pG, of G over the p-adic numbers Z_p, whose unit groups $\mathcal{U}Z_pG$ are nilpotent.

4. SOLVABLE UNIT GROUPS

In this section we shall investigate as to when $\mathcal{U}RG$ is solvable. Of course, a complete answer to this question is dependent on a resolution of Conjecture 1.2 for solvable groups. Bateman [1] gave a necessary and sufficient condition for $\mathcal{U}KG$ to be solvable if $|G| < \infty$ and K is a field of characteristic $p \geq 0$. Bovdi-Khripta [2] and Passman [7] have given alternative characterizations for $p = 2$ or 3 (see also Motose and Ninomiya [1]). We shall prove that if $\mathcal{U}\mathbb{Z}G$ is solvable then the torsion elements of G form an Abelian or a Hamiltonian 2-group T. We shall be able to give necessary and sufficient conditions for $\mathcal{U}\mathbb{Z}G$ to be solvable if G/T is supposed to be nilpotent. We first present the following interesting result of Zassenhaus.

Theorem 4.1. Let G be a solvable group containing a non-normal finite subgroup. Then there is a subgroup S of G containing a normal subgroup N of finite index in S as well as an element $g \in S$ of finite order such that the subgroup generated by g and N is not normal in S.

Proof: Let $G \supset \delta_1 G \supset \delta_2 G \supset \cdots \supset \delta_{k-1} G \supset \delta_k G = 1$ be the derived series of G. By assumption there is an element $g \in G$ of minimal finite order n such that $\langle g \rangle$ is not normal in G. Also there is a natural number $k > 1$ for which $\delta_{k-1} G \neq 1$ but $\delta_k G = 1$. If the theorem is false then there is a counterexample g, G with minimum value of kn.

If the subgroup $\langle g, \delta_{k-1} G \rangle$ is not normal in G then the factor group $\overline{G} = G/\delta_{k-1} G$ forms another counterexample for which

$$\delta_{k-1} \overline{G} = 1 \quad \text{and} \quad o(\overline{g}) \leq n,$$

contradicting the minimum property of g, G. Hence the subgroup $\langle g, \delta_{k-1} G \rangle$ is normal in G.

If $n = n_1 n_2$, $n_1 > 1$, $n_2 > 1$, $(n_1, n_2) = 1$ then $\langle g \rangle = \langle g^{n_1} \rangle \langle g^{n_2} \rangle$ where the order of $\langle g^{n_i} \rangle$ is n_{3-i} $(i = 1,2)$ and at least one of the two factors $\langle g^{n_i}\rangle$ is not normal in G. Hence $n = p^\nu$ for some prime number p. We claim that $\nu = 1$. Clearly due to the choice of g, G $\langle g^p\rangle$ is normal in G. If $\nu > 1$ then $g/\langle g^p\rangle$ is of order p and it generates a subgroup of $G/\langle g^p\rangle$ that is not normal violating the minimum property of the pair g,G. Hence $\nu = 1$ and g is of prime order p.

If $\langle g\rangle$ is not normal in $\langle g,\delta_{k-1}G\rangle$ then $g \notin \delta_{k-1}G$. There is an element x of $\delta_{k-1}G$ not normalizing $\langle g\rangle$. Hence the subgroup $\langle x,g\rangle$ intersects $\delta_{k-1}G$ in an Abelian subgroup A generated by the p conjugates of x under $\langle g\rangle$' such that we have the semi-direct product

$$S = A \rtimes \langle g\rangle, \quad (S:A) = p.$$

S is infinite as otherwise we can simply set N = 1 and G,g is not a counterexample. Since A is finitely generated Abelian it follows that the torsion elements of A form a normal subgroup T(A) of S. If the commutator $(x,g) = x^{-1}g^{-1}xg$ belongs to T(A) then $\langle g\rangle T(A)$ is a finite normal subgroup of S. The cyclic group $\langle x\rangle$ represents the factor group of S over $\langle g,T(A)\rangle$. The centralizer of $\langle g,T(A)\rangle$ in $\langle x\rangle$ is a subgroup $N = \langle x^\mu\rangle$ of finite index μ in $\langle x\rangle$ that forms a normal subgroup of finite index in S such that $\langle g,N\rangle$ is not normal in S contradicting the choice of G. It follows that (x,g) does not belong to T(A). There is a natural number γ prime to p with the property that $(x,g) \notin N = A^\gamma$. On the other hand, N is a normal subgroup of finite index in S such that $\langle g,N\rangle$ is not normal contradicting the choice of G,g. Hence $\langle g\rangle$ is a normal subgroup of $\langle g,\delta_{k-1}G\rangle$. Therefore $\langle g,\delta_{k-1}G\rangle$ is an Abelian normal subgroup of G.

There is an element x of G not normalizing $\langle g\rangle$. The $\langle x\rangle$-conjugates of g generate an elementary Abelian p-group A that is normal in $S = \langle x,g\rangle = \langle x\rangle A$. S is infinite as otherwise we can let N = 1 in contradiction to the choice of G. Therefore x is of infinite order. If A is finite then the subgroup $\langle x\rangle$ represents the factor group of S over A. The centralizer of A in $\langle x\rangle$ is a subgroup $N = \langle x^v\rangle$ of finite index v. Moreover, N is normal of finite index in S such that $\langle g,N\rangle$ is not normal in S in contradiction with the choice of G,g. Hence A is infinite.

It follows that A is the direct product of the conjugate cyclic subgroups $\langle x^{-i}gx^{i}\rangle$ $(i \in \mathbb{Z})$ of prime order p. In this case N, the normal subgroup generated by x^2, is normal of finite index $2p^2$ in S such that $\langle g,N\rangle$ is not normal in S, violating the choice of G,g.

Since there is no counterexample the theorem is demonstrated. □

We shall deduce from the last theorem that, in general, $U\mathbb{Z}G$ contains free subgroups. It would have been possible to deduce several of the results proved earlier from this fact. However since the proofs included are of interest we have chosen not to do so. We shall quote results on free products of groups in proving the next theorem.

Theorem 4.2. Let G be a solvable group containing a non-normal finite subgroup. Then $U\mathbb{Z}G$ contains a free group on two generators.

Proof: By the last theorem we may assume that G contains a normal subgroup N of finite index and an element $g \in G$ such that $\langle g,N\rangle$ is not normal in G. By Wedderburn's structure theorem $Q(G/N)$ is a direct sum of simple rings S_i. Since $\bar{\varepsilon}$ for $\varepsilon = (\Sigma g^i)/o(g)$ is a non-central idempotent of $Q(G/N)$ its projection in at least one S_i is non-central. Thus we have a homomorphism

$$\psi\colon QG \to Q(G/N) \to (D)_n, \quad n > 1$$

to a matrix ring over a division ring. After conjugation we may assume that

$$\psi(\varepsilon) = \left[\begin{array}{c|c} I_r & \\ \hline & \end{array}\right]$$

where I_r is the identity matrix of dimension $r < n$. Let us denote as usual the $n \times n$ matrix with its (i,j) entry one and zeros elsewhere by e_{ij}. Pick $x_{1n}, x_{n1} \in QG$ such that

$$\psi(x_{1n}) = e_{1n}, \quad \psi(x_{n1}) = e_{n1} .$$

Then

$$\psi\big((1-\varepsilon)x_{n1}\varepsilon\big) = e_{n1}, \quad \psi\big(\varepsilon x_{1n}(1-\varepsilon)\big) = e_{1n} .$$

Choose integers λ, μ such that $\lambda\mu > 3$ and set

$$\alpha = \mu(1-\varepsilon)x_{n1}\varepsilon \in \mathbb{Z}G, \quad \beta = \lambda\varepsilon x_{1n}(1-\varepsilon) \in \mathbb{Z}G.$$

Clearly, $\alpha^2 = \beta^2 = 0$, which implies that $1 + \alpha$ and $1 + \beta$ are units of $\mathbb{Z}G$. We have

$$\psi(1+\alpha) = \begin{bmatrix} 1 & & & & \\ & 1 & & & \\ & & \cdot & & \\ & & & \cdot & \\ \mu & & & \cdot & 1 \end{bmatrix}, \quad \psi(1+\beta) = \begin{bmatrix} 1 & & & & \lambda \\ & 1 & & & \\ & & \cdot & & \\ & & & \cdot & \\ & & & \cdot & 1 \end{bmatrix} .$$

Since the map

$$(I_n+e_{n1}\mu) \to (I_2+e_{21}\mu), \quad (I_n+e_{1n}\lambda) \to (I_2+e_{12}\lambda)$$

is a homomorphism, we may follow ψ by this map to get a homomorphism

$$\mathcal{U}\mathbb{Z}G \supset \langle(1+\alpha),(1+\beta)\rangle \overset{\theta}{\to} SL(2,\mathbb{Z})$$

into the 2×2 integral matrices of determinant 1. Replacing e_{12} by $\mu^{-1}e_{12}$ and e_{21} by μe_{21} we have a homomorphism with

$$(1+\alpha) \to X = \begin{bmatrix} 1 & \\ 1 & 1 \end{bmatrix}, \quad (1+\beta) \to Y = \begin{bmatrix} 1 & \delta \\ & 1 \end{bmatrix}, \quad \delta = \lambda\mu > 3.$$

Let us denote by V the group generated by X and Y. It is easy to see that every element of V is of the form

$$\begin{bmatrix} 1 & 0 \\ * & 1 \end{bmatrix} \pmod{\delta}. \tag{4.3}$$

Since $\delta > 3$, it follows that $-I_2 \notin V$. Thus V is a subgroup of $PSL(2,\mathbb{Z}) = SL(2,\mathbb{Z})/\langle -I_2\rangle$. But it is well known (see, for example, Rotman [1], p. 249) that $PSL(2,\mathbb{Z})$ is a free product, $PSL(2,\mathbb{Z}) = \langle A\rangle * \langle B\rangle$ where

$$A = \begin{bmatrix} 0 & 1 \\ -1 & 0 \end{bmatrix}, \quad B = \begin{bmatrix} 0 & 1 \\ -1 & -1 \end{bmatrix}.$$

Therefore, it follows by Kurosh's subgroup theorem (see Kurosh [1], p. 17) that $V = F * V_1 * V_2$ where F is free, V_1 is conjugate to a subgroup of $\langle A\rangle$ and V_2 is conjugate to a subgroup of $\langle B\rangle$. It is easy to check by using (4.3) that no conjugate of A or B belongs to $\pm V$. Thus $V = F$ is free. It follows that $\mathcal{U}\mathbb{Z}G$ contains the free group $\langle (1+\alpha),(1+\beta)\rangle$. □

In order to investigate groups with $\mathbb{Z}G$ solvable we need the following result which is a special case of a theorem of Zassenhaus which characterizes orders with solvable unit groups.

<u>Lemma 4.4</u>. Let $G = K_8 \times \langle x\rangle$ the direct product of the quaternion group of order 8 and a cyclic group of prime order $p > 2$. Then $\mathcal{U}\mathbb{Z}G$ is not solvable.

<u>Proof</u>: See Zassenhaus [3].

<u>Lemma 4.5</u>. Let $G = \langle g_1,\cdots,g_n\rangle$ be a finitely generated group. Then $\gamma_i(G) = \langle (y_1,\cdots,y_i): y_k \in \{g_1,\cdots,g_n\},\ 1 \le k \le i\rangle^G$, the normal subgroup generated by all $(y_1,\cdots,y_i)$, $y_k \in \{g_1,\cdots,g_n\}$.

<u>Proof</u>: We use induction on i and suppose that

$$\gamma_i(G) = \langle (y_1,\cdots,y_i): y_k \in \{g_1,\cdots,g_n\},\ 1 \le k \le i\rangle^G.$$

Then $\gamma_{i+1}(G) = (\gamma_i(G),G)$. Due to (I.3.1) we have to show that for $y_{i+1} \in \{g_1,\cdots,g_n\}$

$$(\gamma_i(G),y_{i+1}) \subseteq H = \langle (y_1,\cdots,y_{i+1}): y_k \in \{g_1,\cdots,g_n\},\ 1 \le k \le i+1\rangle^G.$$

It is enough to prove that for $y = (y_1,\cdots,y_i)$, $g \in G$, $(y^g, y_{i+1}) \in H$. We use the formula

$$(a^b,d) = (a,d)^{(a,b)}(a,b,d) \tag{4.6}$$

which is a consequence of (I.3.1). This implies

$$(a^{bc},d) = (a^b,d)^{(a^b,c)}(a^b,c,d). \tag{4.7}$$

It follows from (4.7) that we may assume $g \in \{g_1,\cdots,g_n\}$. Then (4.6) implies that $(y^g, y_{i+1}) \in H$. □

Theorem 4.8. Suppose that $\mathcal{U}\mathbb{Z}G$ is solvable. Then

(4.9) The torsion elements, $T(G)$, of G form a group which is Abelian or a Hamiltonian 2-group, with every subgroup normal in G.

Conversely, if G is a solvable group satisfying (4.9) and

(4.10) $G/T(G)$ is nilpotent,

then $\mathcal{U}\mathbb{Z}G$ is solvable.

Proof: If $\mathcal{U}\mathbb{Z}G$ is solvable then by (4.2) every torsion subgroup of G is normal and thus the torsion elements form a subgroup $T(G)$ which is Abelian or Hamiltonian. It follows by (4.4) that if $T(G)$ is Hamiltonian then it is a 2-group.

Conversely, let us assume that G is solvable satisfying (4.9) and (4.10). It is enough to prove that

$$\mathcal{U}\mathbb{Z}G = \mathcal{U}(\mathbb{Z}T(G)) \cdot G. \tag{*}$$

Therefore, we may assume that G is finitely generated, say, $G = \langle g_1,\cdots,g_n\rangle$. It follows by (4.5) that if c is the class of nilpotence of $G/T(G)$ then

$$\gamma_{c+1}(G) = \langle (y_1,y_2,\cdots,y_{c+1}) : y_i \in \{g_1,\cdots,g_n\}\rangle^G.$$

This is a finitely generated subgroup of T(G) due to (4.9) and hence is finite. Moreover, since $G/\gamma_{c+1}(G)$ is finitely generated nilpotent it has a finite torsion group. Thus T(G) is finite. It follows by (1.20) that QG has no nilpotent elements and thus every idempotent of QG is central. Now we apply (3.22) to conclude (*).

Corollary 4.11. If G is finite then $\mathcal{U}\mathbb{Z}G$ is solvable if and only if G is Abelian or a Hamiltonian 2-group.

Theorem 4.12. Suppose that $\mathcal{U}QG$ is solvable. Then

(4.13) The torsion elements of G form an Abelian subgroup T(G) with every subgroup normal in G.

Conversely, if G is a solvable group satisfying (4.13) and (4.10) then $\mathcal{U}QG$ is solvable.

Proof: Since $\mathcal{U}QK_8$ is not solvable as seen in (1.14), the "necessity" part of the theorem follows from (4.8). Conversely, let us assume that G is a solvable group, $\delta_k G = \{1\}$, satisfying (4.13) and (4.10). We claim that $\mathcal{U} = \mathcal{U}QG$ is solvable with $\delta_{k+1}\mathcal{U} = \{1\}$. We may suppose that G is finitely generated. It follows as in the last theorem that T(G) is finite and therefore,

$$Q(T(G)) = \sum^{\oplus} F_i$$

a direct sum of fields. Lemma 3.22 implies that every $u \in \mathcal{U}$ can be written as

$$u = \sum f_i g_i, \quad f_i \in F_i, \quad g_i \in G.$$

By using the centrality of idempotents of QG we conclude that for $g \in G$, $F_i^g \subseteq F_i$ and thus

$$\delta_1 \mathcal{U} = (\mathcal{U},\mathcal{U}) \subseteq \sum F_i(G,G),$$

Repeated commutations imply that

$$\delta_k \mathcal{U} \subseteq \sum F_i, \quad \delta_{k+1}\mathcal{U} = \{1\} . \quad \square$$

The following corollary may also be seen directly.

Corollary 4.14. If G is finite $\mathcal{U}QG$ is solvable if and only if G is Abelian.

The counterpart of the next theorem for $p = 2,3$ has been obtained by Bateman [1], Passman [7] and Bovdi-Khripta [2].

Theorem 4.15 (Bateman [1]). If K is a field of positive characteristic $p \neq 2,3$ and G is a finite group then $\mathcal{U}KG$ is solvable if and only if G' is a p-group.

Proof: Let us first suppose that KG is solvable and prove that G' is a p-group. Without loss of generality we may assume that $K = \mathbb{Z}/p\mathbb{Z}$. We use induction on $|G|$. If G contains a normal p-subgroup $P \neq \{1\}$ then by Lemma 3.3 $\mathcal{U}K(G/P)$ is solvable. By induction $(G/P)'$ is a p-group and G' is also p-group. Thus we may assume that G has no nontrivial normal p-subgroup. Let J be the Jacobson radical of KG. Then

$$KG/J \overset{\psi}{\simeq} \sum^{\oplus}(F_i)_{n_i} ,$$

a direct sum of $n_i \times n_i$ matrix rings over finite fields F_i. Lemma 3.3 implies that $\mathcal{U}(KG/J)$ is solvable. It follows (see Rotman [1]) that each $n_i = 1$ and

$$KG/J \overset{\psi}{\simeq} \sum^{\oplus} F_i .$$

Since $G \cap 1 + J$ is a normal p-subgroup by (I.2.21) we may conclude that $G \cap 1 + J = \{1\}$. Therefore,

$$\lambda: G \to \sum^{\oplus} F_i, \quad \lambda(g) = \psi(\overline{g}) \in \sum^{\oplus} F_i$$

is a monomorphism. Thus G is Abelian. The converse is proved in (V.6.16). $\square$

5. GROUP RINGS WITH FC-UNIT GROUPS

In this section we determine groups such that $\mathcal{U}KG$ is an FC-group when $K = \mathbb{Z}$ or a field of characteristic 0.

Lemma 5.1. Let G be a finitely generated group such that, for a commutative ring R, $\mathcal{U}RG$ is an FC-group. Then $\mathcal{U}RG$ is a finite extension of its center $\zeta(\mathcal{U}RG)$.

Proof: Suppose $G = \langle g_1, \cdots, g_\ell \rangle$ and let C_i be the conjugacy class of g_i in $\mathcal{U}RG$. Then $|C_i| < \infty$. Represent $\mathcal{U}RG$ as a group of permutations on C_i. Denoting the kernel by K_i we see that $(\mathcal{U}RG: K_i)$ is finite and hence $(\mathcal{U}RG: \bigcap_1^\ell K_i)$ is finite. We have proved that $(\mathcal{U}RG: \zeta(\mathcal{U}RG))$ is finite.

Let K be a field of characteristic 0 and ζ_i a primitive i-th root of unity contained in an algebraic closure of K. Consider the sequence of fields

$$Q_1 \subset Q_2 \subset \cdots \subset Q_n \subset \cdots, \quad Q_\infty = \cup Q_i, \quad Q_i = Q(\zeta_{p^i}).$$

Then Q_n/Q_1 is cyclic of degree p^{n-1}. Let $\alpha \in Q_\infty$ than $\alpha \in Q_n$ for some n; i.e. $Q_1(\alpha) \subseteq Q_n$. Hence $Q_1(\alpha) = Q_r$ for some $r \le n$. It follows that if we have an intermediate field $Q_1 \subset F \subset Q_\infty$ of finite degree $(F:Q_1)$ then $F = Q_r$ for some r whereas if $(F:Q_1) = \infty$ then $F = Q_\infty$.

Lemma 5.2. (i) If $(K \cap Q_\infty : Q) = s < \infty$ then every cyclotomic polynomial ϕ_n splits into at most s factors in K.

(ii) If $(K \cap Q_\infty : Q) = \infty$ then $K(\zeta_p) \supseteq Q_\infty$.

Proof: Since $(K \cap Q_\infty : Q) = s < \infty$, $K \cap Q_\infty \subseteq Q(\zeta_{p^r})$ for some r and thus $K \cap Q_\infty$ is a Galois extension of Q. If $f(x) \in K[x]$ is an irreducible factor of $\phi_n(x)$ then

$$\phi_n(x) = f^{\sigma_1}(x) \cdot f^{\sigma_2}(x) \cdots f^{\sigma_k}(x), \quad f^{\sigma_1} = f$$

where σ_i are some automorphisms of $K \cap Q_\infty$. Since $k \le s$ we have proved (i).

(ii) Suppose $(K \cap Q_\infty : Q) = \infty$. Then

$$\big((K \cap Q_\infty)(\zeta_p) : Q\big) = \infty, \qquad \big((K \cap Q_\infty)(\zeta_p) : Q_1\big) = \infty$$

and it follows that $(K \cap Q_\infty)(\zeta_p) = Q_\infty$. □

Theorem 5.3 (Sehgal-Zassenhaus [1]). $\mathcal{U}\mathbb{Z}G$ is an FC-group if and only if G is an FC-group and its torsion subgroup T satisfies one of the following:

(i) T is central in G,

(ii) T is Abelian non-central and for $x \in G$ $x^{-1}tx = t^{\delta(x)}$, $\delta(x) = \pm 1$ for all $t \in T$,

(iii) $T = E \times K_8$ where $E^2 = 1$ and K_8 is the quaternion group of order 8. Moreover, E is central in G and conjugation by $x \in G$ induces an inner automorphism on K_8.

Proof: In order to prove the necessity of one of (i) to (iii), we may assume without loss of generality that G is finitely generated. Then due to (5.1) and (II.2.15) either T is Abelian in which case (i) or (ii) holds or $T = A \times E$ in which case (iii) is satisfied.

It remains to prove that if G is an FC-group satisfying one of (i) to (iii) then $\mathcal{U}\mathbb{Z}G$ is an FC-group. We shall consider the three cases separately.

(a) Assume (i). Then every idempotent of QT is central in QG. It follows by Lemma 3.22 that $\mathcal{U}\mathbb{Z}G = (\mathcal{U}\mathbb{Z}T) \cdot G$ and hence is an FC-group.

(b) Assume (ii). We claim that $T = A \times E$ where A is finite, $E^2 = 1$. Suppose for $t \in T$, $g \in G$, $gtg^{-1} = t^{-1}$ then $t^{-1}gt = t^{-2}g$ and thus T can have only a finite number of squares. Therefore, T is of bounded exponent and hence $T = A \times E$ as desired. Now let $\alpha \in \mathcal{U}\mathbb{Z}G$, then it follows by (1.16) that every idempotent of QT is central in QG. Applying Lemma 3.22 we have $\alpha = gu$, $g \in G$,

$u \in \mathcal{U}\mathbb{Z}T$. Writing $\beta \in \mathcal{U}\mathbb{Z}G$ as $\beta = hv$, $h \in G$, $v \in \mathcal{U}\mathbb{Z}T$ we have

$$\beta^{-1}\alpha\beta = \alpha^{\beta} = (gu)^{hv} = (g^h)^v (u^h)^v = g_1^v u^h, \quad g_1 \in G.$$

Since G is an FC-group and u^h takes only two values we have to prove that g_1^v takes only a finite number of values as v runs over $\mathcal{U}\mathbb{Z}T$. In other words, it remains to prove

For a fixed $g \in G$ there are only a finite number of distinct $u^{-1}gu$ for $u \in \mathcal{U}\mathbb{Z}T$. (*)

For any $\alpha = \Sigma\alpha_g g \in \mathbb{Z}G$ write $\alpha^* = \Sigma\alpha_g g^{-1}$ then

$$g^u = u^{-1}gu = u^{-1}gug^{-1}g = \gamma g \quad \text{where} \quad \gamma = 1 \text{ or } u^{-1}u^*.$$

Since γ satisfies $\gamma^* = \gamma^{-1}$, we have $\gamma = \pm t$, $t \in T$. But since γ has augmentation one, $\gamma = t$. Hence,

$$g^u = tg, \quad t = a\tau, \quad a \in A, \quad \tau \in E.$$

We claim that $\tau = 1$. To prove this we may take E to be finite and let $\tau \neq 1$. We have

$$QE = \overset{\oplus}{\sum} Q = \overset{\oplus}{\sum} e_i QE, \quad \text{where} \quad 1 = \sum e_i.$$

Then, because $\tau \neq 1$, we can renumber the i's so that

$$\tau = \sum \tau e_i = -e_1 + \sum_{i \neq 1} \tau e_i .$$

Also, we have a decomposition of QA,

$$QA = eQA \oplus \cdots$$

where $e = \frac{1}{|A|}\Sigma_{a\in A}\, a$. Now, $Q(A \times E)$ has an idempotent ee_1 with

$$ee_1 Q(A \times E) = ee_1(QE)A = e(e_1 QE)A \simeq Q$$

with $t(ee_1) = \tau(ee_1) = -ee_1$. But, due to centrality of idempotents,

$$tee_1 = u^{-1}gug^{-1}ee_1 = u^{-1}guee_1 g^{-1} = ee_1 \quad \text{as} \quad uee_1 \in Q.$$

This is a contradiction. Thus $\tau = 1$ and $g^u = ag$, $a \in A$ which proves (*) because A is finite.

(c) Assume (iii). In this case, $T = E \times K_8$ and it follows by (1.24) that QG has no nilpotent elements. Thus by (1.8) every idempotent of QG is central which implies by (3.22) that $\mathcal{U}\mathbb{Z}G = (\mathcal{U}\mathbb{Z}T) \cdot G$. But by (II.2.5) we know that $\mathcal{U}\mathbb{Z}T = \pm T$. Hence $\mathcal{U}\mathbb{Z}G = \pm G$ and it follows that $\mathcal{U}\mathbb{Z}G$ is an FC-group. □

Theorem 5.4 (Sehgal-Zassenhaus [1]). Let K be a field of characteristic 0. Then $\mathcal{U}KG$ is an FC-group if and only if one of the following holds:

(i) G is Abelian;

(ii) G is a non-Abelian FC-group whose torsion subgroup T is finite central;

(iii) G is a non-Abelian FC-group with central torsion subgroup

$$T = Z(p^\infty) \times B, \quad |B| < \infty$$

for some prime p, such that the derived subgroup G' is contained in $Z(p^\infty)$. Moreover, if Q_∞ is the field obtained by adjoining all p^i-th roots of unity to Q then $(Q_\infty \cap K: Q) = s < \infty$.

5.5. Proof of "Necessity"

Suppose that $\mathcal{U}KG$ is an FC-group. Then so are $\mathcal{U}QG$ and $\mathcal{U}\mathbb{Z}G$. It follows by the last theorem that every subgroup of T is normal in G and T is Abelian or Hamiltonian. We know (1.14) that

$$QK_8 = Q \oplus Q \oplus Q \oplus Q \oplus S$$

where $S = \{Q + Qi + Qj + Qk\}$ is the division algebra of the rational quaternions. No power of S is contained in its center, as seen, for example, by computing $(1+ni)^n$. This implies by (5.1) that $\mathcal{U}QK_8$ is not an FC-group. Hence T is Abelian. In order to prove that T is central we may assume that $G = \langle t,x \rangle$, $t \in T$, $x \notin T$. We have

$$QT = \sum^{\oplus} F_i$$

a direct sum of cyclotomic fields. Moreover, since each subgroup of T is normal in G each idempotent of QT is central in QG (1.16). Thus $x^{-1}F_i x = F_i$ for each i. Let $\overline{F}_i$ be the fixed fiel of F_i under conjugation by x. Then since by (5.1),

$$\{\mathcal{U}QG : \zeta(\mathcal{U}QG)\} = n,$$

for some $n \geq 1$, we have

$$f \in F_i \Rightarrow f^n \in \overline{F}_i \ .$$

Suppose there exists a $\beta \in F_i$, $\beta \notin \overline{F}_i$. Then for all $\lambda \in Q$

$$(1+\lambda\beta)^n = d_\lambda \in \overline{F}_i \ .$$

Taking distinct nonzero values λ_j, $1 \leq j \leq n$, for λ we have

$$n\lambda_j\beta + \binom{n}{2}(\lambda_j\beta)^2 + \cdots + (\lambda_j\beta)^n = c_j \in \overline{F}_i \ .$$

This is a set of n non-homogeneous linear equations

$$x_1\lambda_j + x_2\lambda_j^2 + \cdots + x_n\lambda_j^n = c_j, \quad 1 \leq j \leq n,$$

over $\overline{F}_i$, in the variables $x_1,x_2,\cdots,x_n$, with the determinant

$$\begin{vmatrix} \lambda_1 & \lambda_1^2 & \cdots & \lambda_1^n \\ \lambda_2 & \lambda_2^2 & \cdots & \lambda_2^n \\ \vdots & & & \\ \lambda_n & \lambda_n^2 & \cdots & \lambda_n^n \end{vmatrix} \neq 0.$$

Its solution $n\beta$, $\binom{n}{2}\beta^2,\cdots,\beta^n$ therefore belongs to $\overline{F}_i$, implyin that $\beta \in \overline{F}_i$. This is a contradiction proving that $\overline{F}_i = F_i$. We have proved that x commutes elementwise with F_i and hence with t. Thus T is central in G.

We wish to prove that if T is infinite and G is non-commut tive then (iii) holds. Let us choose $1 \neq (g,x) = t \in T$, $g,x \in G$. We assert:

(5.6) G' is contained in any infinite subgroup H of T.

If $t \notin H$ then there is an infinite chain

$$H_1 \subsetneq H_2 \subsetneq H_3 \subseteq \cdots$$

of finite subgroups of H such that $t \notin H_i$ for any i. Write

$$e_i = \hat{H}_i/|H_i|, \quad \hat{H}_i = \sum_{h \in H_i} h.$$

Then $(1-t)e_i = (1-t)e_j \iff e_i = e_j$. We have the units

$$u_i = e_i x + (1-e_i), \quad u_i^{-1} = e_i x^{-1} + (1-e_i).$$

This gives us distinct conjugates of g,

$$g^{u_i} = u_i^{-1} g u_i = \big(e_i t + (1-e_i)\big) g$$

because

$$g^{u_i} = g^{u_j} \iff (1-t)e_i = (1-t)e_j \iff e_i = e_j \iff u_i = u_j .$$

Thus $t \in H$ and $G' \subseteq H$, proving (5.6).

Now, T is not a reduced group as in that case it contains an infinite direct sum of finite subgroups (see Fuchs [1], p. 117) in contradiction with (5.6). Thus T contains a divisible subgroup and again (5.6) implies that

$$T = Z(p^\infty) \times B, \quad |B| < \infty, \quad G' \subseteq Z(p^\infty),$$

for some prime p.

It remains to prove that if $Q_\infty = \cup Q_i$, $Q_i = Q(\zeta_{p^i})$ then $(K \cap Q_\infty : Q) < \infty$. If not, then by (5.2) we have $K(\zeta_p) \supseteq Q_\infty$. Let $1 \neq (g,x) = t \in Z(p^\infty)$ be of order p^m. We shall produce p^ℓ conjugates of g for an arbitrary ℓ. Let $y^{p^\ell} = t$, $y \in Z(p^\infty)$. Then

$$K\langle y\rangle = \hat{t}\, K\langle y\rangle \oplus (1-t)K\langle y\rangle, \quad \hat{t} = (\Sigma t^i)o(t).$$

It is easy to see by counting that $\hat{t}\, K\langle y\rangle$ has K-dimension p^ℓ

which implies that $(1-t)K\langle y\rangle$ has K-dimension $p^{\ell+m} - p^{\ell}$. But

$$K\langle y\rangle = K \oplus K(\zeta_p) \oplus \cdots \oplus K(\zeta_p).$$

Thus $(1-t)K\langle y\rangle$ has at least p^{ℓ} primitive idempotents $\{e_i\}$ giving rise to distinct conjugates of g

$$g^{u_i} = u_i^{-1} g u_i, \quad u_i = e_i x + (1-e_i).$$

This is a contradiction proving $(K \cap Q_\infty : Q) < \infty$ and completing the proof of the "necessity" part of Theorem 5.4.

5.7. Proof of "Sufficiency"

We have to prove that each of (ii) and (iii) implies that KG is an FC-group. We consider the two cases separately.

(a) Assume (ii). Then

$$KT = \sum^{\oplus} F_i, \quad 1 \le i \le n,$$

a direct sum of fields and by Lemma 3.22 every unit of KG can be written as

$$u = \sum_{1}^{n} f_i g_i, \quad f_i \in F_i, \quad g_i \in G$$

and thus has a finite number of conjugates.

(b) Assume (iii). We need the next result.

Lemma 5.8. Let $y,t \in Z(p^\infty)$, $o(t) = p^m$, $y^{p^\ell} = t$. Then $(1-t)\varepsilon_i$. for at most $a = s\,m\,p^m|B|$ primitive orthogonal idempotents ε_i of $K(\langle y\rangle \times B)$.

Proof: We have

$$K(\langle y\rangle \times B) = \hat{t}\; K(\langle y\rangle \times B) \oplus (1-t)K(\langle y\rangle \times B)$$

$$K\langle t\rangle = \hat{t}\; K\langle t\rangle \oplus (1-t)K\langle t\rangle, \quad \hat{t} = \left(\textstyle\sum t^i\right)/o(t).$$

Also,

$$Q\langle y\rangle = \sum_{1 \le i \le \ell+m}^{\oplus} Q(\xi_i)$$

where ξ_i is a p^i-th root of unity with

$$1 = \sum e_i, \quad y = \sum \xi_i, \quad t - 1 = \sum(\xi_i^{p^\ell} - 1).$$

Since $(t-1)e_i = 0 \iff \ell \geq i$, we can deduce that $(t-1)e_i \neq 0$ for m values of i. Thus every primitive idempotent e of $(1-t)Q<t>$ splits into at most m primitive orthogonal idempotents of $Q<y>$. It follows that e splits into a sum of no more than $m|B|$ primitive orthogonal idempotents of $Q(<y> \times B) = A$. Now, a primitive idempotent ε of A gives

$$\varepsilon A = Q[x]/\langle\phi_n(x)\rangle$$

where $\phi_n(x)$ is a cyclotomic polynomial. By Lemma 5.2, $\phi_n(x)$ splits into at most s irreducible factors in $K[x]$ and thus ε splits into at most s primitive orthogonal idempotents of

$$K(<y> \times B) = K \otimes A = (K \otimes \Sigma\varepsilon A) = \Sigma K \otimes \varepsilon A.$$

It follows that e splits into a sum of no more than $sm|B|$ primitive orthogonal idempotents of $K(<y> \times B)$. Since there are (p^m-1) possibilities for e, the proof of the lemma is complete.

Let $u \in UKG$. We wish to prove that u has a finite number of conjugates. The torsion subgroup, $T\langle supp(u)\rangle$, of the group generated by the support of u is contained in $<t_1> \times B$ for some $t_1 \in Z(p^\infty)$. Writing

$$K(<t_1> \times B) = \sum_{1 \leq i \leq r}^{\oplus} F_i,$$

a direct sum of fields we may express u by (3.22) as

$$u = \sum_1^r f_i g_i, \quad f_i \in F_i, \quad g_i \in G.$$

Since G is an FC-group, the group $G_1 = \langle supp(u), (g_i,G), 1 \leq i \leq r\rangle$ is finitely generated and has finite torsion group

$$T(G_1) \subseteq <t> \times B, \quad o(t) = p^m$$

for some $t \in Z(p^\infty)$. We claim that u has at most $p^m a$ conjugates where a, depending on $\langle t\rangle \times B$ and K only, is defined in (5.8). If not, suppose we have $(p^m a+1)$ conjugates u^{v_i} of u. Let

$$T\langle G_1, v_i, 1 \le i \le p^m a+1\rangle \subseteq \langle y\rangle \times B.$$

We have the direct sum of fields

$$K(\langle y\rangle \times B) = \sum^{\oplus} K_i, \quad 1 = \sum \varepsilon_i.$$

Write

$$u = \sum \alpha_i \ell_i, \quad v = v_j = \sum \beta_i h_i, \quad \alpha_i, \beta_i \in K_i, \quad \ell_i, h_i \in G$$

where $\{\ell_i\} = \{g_i\}$. Then

$$u^v = \sum \alpha_i \ell^{h_i} = \sum \alpha_i \ell_i t_i, \quad t_i = t^{b_i} \text{ for some integer } b_i,\ 0 \le b_i < p^m.$$

Notice that

$$\alpha_i \ell_i t_i \ne \alpha_i \ell_i \Rightarrow \varepsilon_i(1-t_i) \ne 0 \Rightarrow \varepsilon_i(1-t) \ne 0.$$

But by (5.8) there exist at most a values of i for which $\varepsilon_i(1-t) \ne 0$. Thus u^v has no more than ap^m possibilities. This proves our assertion and completes the proof of the theorem. □

6. NORMAL SUBGROUPS OF $\mathcal{U}((\mathbb{Z}/n\mathbb{Z})G)$

Let G be a finite group and $K = \mathbb{Z}/n\mathbb{Z}$. In this section we characterize all subgroups H of G which are normal in $\mathcal{U}KG$. As a consequence we can give necessary and sufficient conditions for $G \lhd \mathcal{U}KG$ and for every unit of KG to be trivial. The results presented are due to Pearson ([1],[2]) and Bovdi and Khripta [1]. The last paper discusses these questions when G is not assumed to be finite. The special case when G is a p-group is due to Eldridge [1 Analogues of these results for integral group rings have been discussed in Chapter II. Denote by S_3 and A_3 the symmetric and alternating groups on 3 letters.

Theorem 6.1 (Bovdi-Khripta-Pearson). Let $(\mathbb{Z}/n\mathbb{Z})G$ be the group ring of a finite group G over $\mathbb{Z}/n\mathbb{Z}$, the integers modulo n. Then a subgroup H of G is normal in $\mathcal{U}((\mathbb{Z}/n\mathbb{Z})G)$ if and only if H is contained in the center of G or $n = 2$, $G \simeq S_3$ and $H \simeq S_3$ or A_3.

The proof, presented here, is a combination of arguments of Pearson and Bovdi and Khripta. The proof of the "sufficiency" is given in (6.2) and that of the "necessity" part is broken into several lemmas.

6.2. Proof of "Sufficiency"

We shall prove that $S_3 \lhd \mathcal{U} = \mathcal{U}((\mathbb{Z}/2\mathbb{Z})S_3)$. Since A_3 is the sylow 3-subgroup of S_3 it follows that $A_3 \lhd \mathcal{U}$. Write

$$S_3 = \langle a,b\ \ a^2 = b^3 = 1,\ \ ba = ab^2\rangle .$$

Then

$$\theta\colon S_3 \to \mathbb{Z}/2\mathbb{Z} \oplus M_2(\mathbb{Z}/2\mathbb{Z})$$

given by

$$\theta(a) = \left(1, \begin{bmatrix} 1 & 1 \\ 0 & 1 \end{bmatrix}\right), \quad \theta(b) = \left(1, \begin{bmatrix} 0 & 1 \\ 1 & 1 \end{bmatrix}\right)$$

defines a group homomorphism and so can be extended to a ring homomorphism

$$\theta\colon (\mathbb{Z}/2\mathbb{Z})S_3 \to \mathbb{Z}/2\mathbb{Z} \oplus M_2(\mathbb{Z}/2\mathbb{Z}).$$

Here, $M_2(R)$ denotes the ring of 2×2 matrices over R. It is easy to check that the kernel of θ is $\{0,\gamma\}$ where $\gamma = 1 + b + b^2 + a + ab + ab^2$. Thus θ is onto with nilpotent kernel $J = \{0,\gamma\}$. We have by (3.3) an induced epimorphism

$$\mathcal{U}((\mathbb{Z}/2\mathbb{Z})S_3) \to \mathcal{U}M_2(\mathbb{Z}/2\mathbb{Z})$$

with kernel $1 + J$. Since $\mathcal{U}M_2(\mathbb{Z}/2\mathbb{Z})$ has order 6 and θ is one to one when restricted to S_3 it follows that

$$\mathcal{U}((\mathbb{Z}/2\mathbb{Z})S_3) = S_3(1+J).$$

The centrality of γ implies that S_3 is normal in $\mathcal{U}$. □

Now we prove the "necessity" part of the theorem. Throughout this section we shall let G be the given finite group with a non-central subgroup H. K will be $\mathbb{Z}/n\mathbb{Z}$ and $\mathcal{U}$ will be the unit group of KG with $H \triangleleft \mathcal{U}$.

Lemma 6.3. K has no non-trivial idempotent or nilpotent elements.

Proof: (i) Suppose $e = e^2 \in K$, $e \neq 0,1$. Then for any $u \in \mathcal{U}KG$, $e + u(1-e) \in \mathcal{U}KG$. Thus, for any $h \in H$ there exists an $h' \in H$ such that

$$\big(e + u(1-e)\big)h = h'\big(e + u(1-e)\big).$$

Multiplying by e it follows that $eh = h'e$ so that $h = h'$. Therefore,

$$u(1-e)h = hu(1-e)$$

which implies $uh = hu$. This is a contradiction as H is not centr

(ii) Suppose that $0 \neq \alpha \in K$ is nilpotent. Then for $g \in G$, $g - \alpha \in \mathcal{U}KG$. Thus for any $h \in H$, there exists an $h' \in H$ such that

$$(g-\alpha)h = h'(g-\alpha).$$

It follows that $h = h'$ and $gh = hg$. Hence H is central. This is a contradiction, proving the lemma. □

We may assume from now on that K is the prime field of p elements.

Lemma 6.4. Let $K_8 = \langle i,j\colon i^2 = j^2 = t,\ t^2 = 1,\ ji = ijt\rangle$ be the quaternion group of order 8. Suppose that F is a prime field of p elements with $p \neq 2$. Then K_8 is not normal in $\mathcal{U}FK_8$.

Proof: We know from (1.13) that for $\gamma = (ai+bj+cij)(1-t) \in FK_8$, $\gamma^2 = 0$ if and only if $a^2 + b^2 + c^2 = 0$. Since each of the sets

$$X = \left\{x^2\colon x = 0,1,2,\cdots,\tfrac{p-1}{2}\right\},\quad Y = \left\{-(1+y^2)\colon y = 0,1,\cdots,\tfrac{p-1}{2}\right\}$$

has $(p+1)/2$ distinct elements it follows that $x^2 + y^2 = -1$ for some $0 \le x,y \le (p-1)/2$. So let us pick $a = 1$ and b, c such that $a^2 + b^2 + c^2 = 0$. Set $\gamma = (ai + bj + cij)(1-t)$. By direct computation,

$$(1-\gamma)i(1+\gamma) = i(1+4a^2) + ij(2b+4ac) + tj(2c-4ab) + tij(-2b-4ac) + j(-2c+4ab) + it(-4a^2).$$

Thus if $p \neq 5$, $1 + 4a^2 \neq 0$ and $(1-\gamma)i(1+\gamma) \notin K_8$. If $p = 5$, take $b = 1$ then $2c \neq 4ab$ as otherwise $c = 2a$ and $a^2 + b^2 + c^2 = a^2 + 1 + 4a^2 = 1 \neq 0$. Again, $(1-\gamma)i(1+\gamma) \notin K_8$.

<u>Lemma 6.5</u>. (i) If $n \neq 2$ then H is Abelian with every subgroup normal in G.

(ii) If $n = 2$ then the elements of odd order in H form an Abelian group with every subgroup normal in G.

<u>Proof</u>: Let $h \in H$ with $h^m = 1$. Then for $x \in G$ and

$$\gamma = (1 + h + \cdots + h^{m-1})x(1-h)$$

we have $\gamma^2 = 0$ and $(1+\gamma) \in \mathcal{U}KH$. Therefore,

$$\begin{aligned}(1+\gamma)^{-1}h(1+\gamma) &= (1-\gamma)h(1+\gamma)\\ &= [1 - (\Sigma h^i)x(1-h)]h[1 + (\Sigma h^i)x(1-h)]\\ &= [1 - (\Sigma h^i)x(1-h)]h + (\Sigma h^i)x(1-h)\\ &= h - (\Sigma h^i)x(h-h^2-1+h)\\ &= h + (\Sigma h^i)xh^2 - 2(\Sigma h^i)xh + (\Sigma h^i)x \in H.\end{aligned}$$

Since $n \neq 2$ we have

either $h^i xh = h$	or $h^i xh = h^j x$ for some j	or $h^i xh = h^k xh^2$ for some k
$\Downarrow$	$\Downarrow$	$\Downarrow$
$x \in \langle h \rangle$	$h^{i-j}x = xh^{-1}$	$h^{i-k}x = xh$
$\Downarrow$	$\Downarrow$	$\Downarrow$
$\langle h \rangle \lhd G$	$\langle h \rangle \lhd G$	$\langle h \rangle \lhd G$.

In any case H is Abelian or Hamiltonian. But by the last lemma

K_8 is not a subgroup of H. It follows that H is Abelian.

(ii) n = 2. Suppose that m > 2. From the computation above we have

$$(1+\gamma)^{-1}h(1+\gamma) = h + (\Sigma h^i)xh^2 + (\Sigma h^i)x \in H.$$

It follows that for some i and j

$$h^i x = h^j xh^2 \Rightarrow h^{i-j} = xh^2x^{-1} \Rightarrow \langle h^2\rangle \triangleleft G.$$

We have proved that the odd order elements of H form an Abelian subgroup as desired. □

Lemma 6.6. $p \nmid |H| \Rightarrow H$ is cyclic of prime power order.

Proof: By the last lemma H is Abelian. Suppose that H contains a direct product of cyclic groups, $H \supseteq \langle a\rangle \times \langle b\rangle$. Again, by the last lemma, $\langle a\rangle \triangleleft G$, $\langle b\rangle \triangleleft G$. This implies that

$$e = (\Sigma a^i)/o(a), \quad f = (\Sigma b^i)/o(b)$$

are central idempotents. Therefore, for $h \in H$, $g \in G$

$$\left(e - g^{-1}(1-e)\right)h\left(e + g(1-e)\right) = h' \in H.$$

Thus $eh - (g^{-1}hg)(1-e) = h'$. It follows that $eh = eh'$ and $h' = ha^j$ for some j. Then $e - (h,g)(1-e) = a^j$ and $(h,g) \in \langle a\rangle$. Similarly, we can conclude that $(h,g) \in \langle b\rangle$. This implies that $(h,g) = 1$ and H is central. □

Lemma 6.7. If $\mathcal{U}M_n(F)$ has a non-central normal cyclic subgroup for a field F then $|F| = 2$, $n = 2$.

Proof: See Artin [1], page 165.

Lemma 6.8. $p \nmid |H| \Rightarrow p = 2$, $|H| = 3$, $6 \mid |G|$.

Proof: We know by (6.6) that $H = \langle a\rangle$. Let us denote by J the Jacobson radical of KG. Then, by Wedderburn's theorem,

$$KG/J = \sum^{\oplus}(F_i)_{n_i}, \qquad (*)$$

a direct sum of full matrix rings $(F_i)_{n_i}$ over finite fields F_i. In view of (3.3), the natural projection $KG \to KG/J$ induces an epimorphism

$$\psi\colon UKG \to U(KG/J).$$

The kernel of ψ is $1+J$ which is a p-group. Thus $\psi H \simeq H$ as H has no p-element. Suppose that

$$\psi(a) = (a_1,\cdots,a_t).$$

Since $1+J$ is a normal p-group and H is a normal p'-group, it is easy to see that not every a_i is central. Without loss of generality we may assume that a_1 is not central in $U((F_1)_{n_1})$. But $\langle a_1\rangle \lhd U((F_1)_{n_1})$. It follows by (6.7) that

$$n_1 = 2, \quad |F_1| = 2, \quad U((F_1)_{n_1}) = S_3.$$

Write $S_3 = \langle (a_1,1,\cdots,1),(b,1,1,\cdots,1) = b\rangle$. Then ψH contains the element

$$\psi(a)^{-1}b^{-1}\psi(a)b = (a_1,1,\cdots,1).$$

It follows that $(1,a_2,\cdots,a_t) \in \psi(H)$. Because H is cyclic of prime power order we can conclude that $a_2 = a_3 = \cdots = a_t = 1$. Thus H is cyclic of order 3, $|K| = 2$. Since H is a non-central normal subgroup of order 3, G must have an element of order 2. □

Lemma 6.9. $|K| = 2$, $g \in G$, $g \notin H$, $o(g) = 2^m$, $m > 1 \Rightarrow g$ centralizes H.

Proof: Clearly $(1+g+g^2) \in UKG$. Therefore given $h \in H$ there exists $h' \in H$ such that

$$(1+g+g^2)h = h'(1+g+g^2).$$

Comparing elements in H we conclude that $h = h'$. Suppose that g does not centralize H then $gh = hg^2$, $g^2h = hg$. Thus we get

$$h^{-1}gh = g^2, \quad h^{-1}g^2h = g \quad \text{and} \quad g^4 = g.$$

This is a contradiction, proving the result. □

The next two propositions will complete the proof of Theorem 6.1.

<u>Proposition 6.10</u>. $p \nmid |H| \Rightarrow p = 2, \quad H \simeq A_3, \quad G \simeq S_3$.

<u>Proof</u>: By (6.8) $p = 2$ and $H = \langle a \rangle$ is cyclic of order 3. Moreover,

(6.11) Every element of odd order centralizes H.

Therefore, by the last two results, there exists an element $z \in G$ of order 2 such that $\langle a,z \rangle \simeq S_3$. We claim

(6.12) $4 \nmid |G|$.

If not, we can find a subgroup M of order 4 containing z. In view of (6.9) this subgroup can't be cyclic. Therefore, $M = \langle z,y \colon z^2 = y^2 = 1 \rangle$. This gives us a unit $1 + y + z$. Thus

$$(1+y+z)a = \alpha(1+y+z) \quad \text{with} \quad \alpha = a \quad \text{or} \quad a^2.$$

Comparing elements in H we get $\alpha = a$. This implies that either $ya = ay$, $za = az$ or $ya = az$, $y \in \langle z,a \rangle$. Each of the conclusions is a contradiction, proving that $4 \nmid |G|$.

We assert

(6.13) $g \in G, \quad o(g) = \text{odd} \Rightarrow g \in H$.

If not, suppose there is a $g \in G$, $g \notin H$ of odd order. Then by (6.5), $ga = ag$ and

either $zgz^{-1} = g^2 \neq g$	or $zgz^{-1} = g$
$\Downarrow$	$\Downarrow$
$\eta = (1+z)(g+g^z)$ satisfies $\eta^2 = 0$	$\eta = (1+z)(a+a^2)g$ satisfies $\eta^2 = 0$
$\Downarrow$	$\Downarrow$
$(1+\eta)a = a(1+\eta)$	$(1+\eta)a = a(1+\eta)$
$\Downarrow$	$\Downarrow$
$az(g+g^z) = z(g+g^z)$	$z(1+a^2) = z(1+a)$.

Either consequence is a contradiction proving (6.13). Thus G/H is a 2-group. It follows by (6.12) that $(G:H) = 2$ and $G \simeq S_3$. □

Proposition 6.14. $p \,\Big|\, |H| \Rightarrow p = 2, \quad G = H \simeq S_3$.

Proof: We shall first prove that $p = 2$. If not, let $p \geq 3$. Pick $x \in H$, $o(x) = p$, $g \in G$, $h \in H$. Since by (6.5) H is Abelian, its Sylow p-subgroup is normal and $(1-x)g$ is nilpotent. So there exists an $x' \in H$ such that

$$\big(1 + (1-x)g\big)x = x'\big(1 + (1-x)g\big), \quad x + gx + x'xg = xgx + x' + x'g.$$

It follows that $x = x'$ and therefore $gx = xg$. This is a contradiction, proving that $p = 2$.

Suppose H is a 2-group. Then by considering the equation

$$(1+h_1+h_2)h_1(1+h_1+h_2)^{-1} \in H$$

for $h_1, h_2 \in H$ we can conclude that H is commutative and thus $G \neq H$. Then pick $g \in G$, $g \notin H$. Given $h \in H$, there exists $h' \in H$ such that

$$\big(1 + (1+h)g\big)h = h'\big(1 + (1+h)g\big).$$

It is easily seen that $h = h'$. Therefore,

$$(1+h)gh = h(1+h)g \Rightarrow gh + hgh = hg + h^2g \Rightarrow gh = hg.$$

This is a contradiction, which proves that a 2-group cannot be normal in $\mathcal{U}$. By (6.5) the odd order elements form a normal subgroup M of H. Clearly, M is normal in $\mathcal{U}$. M is not central as otherwise $H = M \times T$ with T a normal 2-subgroup of $\mathcal{U}$. It follows by (6.10) that $M \simeq A_3$ and $G \simeq S_3$. Hence, $H = G$. □

Theorem 6.15 (Pearson [1]). Let $K = \mathbb{Z}/n\mathbb{Z}$ and let $G \neq \{1\}$ be a finite group. Then all the units of KG are trivial if and only if $(n = 2, |G| = 2,3)$ or $(n = 3, |G| = 2)$.

Proof: Since $G \lhd \mathcal{U}KG$ we have by the last theorem that G is Abelian or $G \simeq S_3$, $n = 2$. The second possibility doesn't arise as $\Sigma_{g \in S_3} g = \eta$ is a nilpotent element giving rise to a nontrivial unit $1 + \eta$. Thus G is Abelian. At first suppose every prime dividing $|G|$ divides n. Then $\Sigma_{g \in G} = \eta$ is a nilpotent element. Therefore,

$$1 + \eta = 1 + \sum_{g \in G} g = ah, \quad h \in G, \quad a \in K.$$

This can only happen if $|G| = 2$ and $n = 2$.

The remaining possibility is that there exists a prime q such that $q \,\big|\, |G|$, $q \nmid n$. Let H be a subgroup of G of order q,. then $e = (\Sigma_{h \in H} h)/q$ is an idempotent. If there is a $g \in G$, $g \notin$ H then $\big(e - (1-e)g\big)$ is a nontrivial unit of KG. Thus we have $|G| =$ and $q \nmid n$.

Notice that n is square free. Otherwise we shall have a nontrivial unit of the form $\frac{n}{a} g + 1$, $g \in G$ for a suitable divisor a of n. Also, if $n = p_1 p_2 \cdots p_s$ a product of distinct primes than

$$KG = K_1 G \oplus \cdots \oplus K_s G, \quad K_i = \mathbb{Z}/p_i\mathbb{Z}.$$

It follows that all units of $K_i G$ are trivial. Thus it is enough to prove that if all units of KG are trivial then

(6.16) $\quad n = p,\ |G| = q,\ q \neq p \Rightarrow n = 2,\ q = 3 \ \text{ or } \ n = 3,\ q = 2.$

We know (II.2.6) that

$$KG = F_1 \oplus \cdots \oplus F_\ell,$$

a direct sum of fields. Counting units we have an equation of the form

$$(p-1)q = (p-1) \cdots (p^m - 1).$$

This implies either $p = 2$ or $(p = 3,\ q = 2)$. It remains to consid the possibility $p = 2$, $|G| = q$. This time we get

$$KG = K \oplus K(\xi), \quad \xi^q = 1,$$

which implies by counting the number of elements that $2^q = 2(q+1)$. This can happen only if $q = 3$. The proof of the necessity of the conditions is complete. The "sufficiency" part follows trivially by using the formula (II.2.6). □

CHAPTER VII

RESEARCH PROBLEMS

There are numerous unsolved problems in the area of group rings. Some of these can be found in Bovdi [3], Dennis [2], Passman [3],[5] and Zalesskii and Mikhalev [1]. We shall list below problems relevant to the material covered in this book. Some of these are of the well known variety and no doubt some will turn out to be easy. They are listed, roughly, in the order they are encountered in this book. Throughout, KG will denote a group algebra over a field K and RG will be a group ring over a commutative unital ring R such that whenever G has an element of prime order p, p is a non-unit of R. We begin with (I.2.10).

Problem 1. If G is torsion free then KG has no nontrivial idempotents.

Related to this is (I.2.11). Let K have characteristic 0 then we know

$$e = \sum e(g)g = e^2 \in KG \Rightarrow e(1) = r/s, \quad (r,s) = 1, \quad r,s \in \mathbb{Z}.$$

Problem 2. For every prime p dividing s there is an element of order p in G.

This is known to be true for polycyclic by finite groups (see (I.2.15)). The last two problems are special cases of (I.2.18) whic we state next.

Problem 3. If R has no nontrivial idempotents then RG has no nontrivial idempotents.

This is known to hold for finite groups or supersolvable groups and for polycyclic by finite groups if R has characteristic 0 (see Chapter I). Formanek [2] solved Problem 1 for Noetherian group if K has characteristic 0 (see (I.2.9)). So we list separately the special case of characteristic $p > 0$.

Problem 4. If K has characteristic $p > 0$ and G is torsion fre Noetherian then KG has no nontrivial idempotents.

An affirmative answer to the next question, of course, will settle this if $K = \mathbb{Z}/p\mathbb{Z}$.

Problem 5. Can every idempotent of $(\mathbb{Z}/p\mathbb{Z})G$ be lifted to $\mathbb{Z}_pG$?

Incidentally, an easy argument for the case of G supersolvabl for Problem 4 is as follows:

$$e = e^2 \in KG \Rightarrow \tilde{e}(x) = \sum_{g\sim x} e(g) = 0 \quad \text{for all} \quad x \in G.$$

This implies that e or $e - 1 \in \Delta(G,G')$ and therefore

$$e \quad \text{or} \quad e - 1 \in \bigcap_n \Delta(G,G')^n = \Big(\bigcap_n \Delta^n(G')\Big) KG = 0.$$

But since G' is torsion free nilpotent (I.3.2), $\bigcap_n \Delta^n(G') = 0$ (se Jennings [3] and Formanek [1]). Thus $e = 0$ or 1.

Let K have characteristic 0 and $e = \Sigma e(g)g = e^2 \in KG$. The Kaplansky proved that $e \neq 0,1 \Rightarrow 0 < e(1) < 1$ and Zalesski [1] proved that $e(1) \in Q$. Bass [2] extended this to prove that $E = Q(\tilde{e}(g): g \in G)$ is an Abelian normal extension of Q. He furthe made the following conjecture which he supported with a proof for th polycyclic by finite case.

Problem 6. E is contained in $Q(\xi_\ell)$ where ξ_ℓ is a primitive ℓ-th root of unity and ℓ is the least common multiple of finite orders $o(g)$ of the group elements $g \in G$ with $\tilde{e}(g) \neq 0$.

Perhaps there exists an analogue for $e(g)$ of the Kaplansky's result.

Problem 7. If g has a finite number h of conjugates then $|\tilde{e}(g)| < h$.

Problem 8. Is there an analogue of Problem 2 for $\tilde{e}(g)$?

The next problem is (II.1.5) which is known to be true if G is polycyclic by finite or if no rational prime is a unit in R.

Problem 9. If R is an integral domain of characteristic 0,

$$\alpha \in RG, \quad \alpha^n = 1, \quad \alpha(1) \neq 0 \Rightarrow \alpha = \alpha(1).$$

Then there is the well known

Problem 10. If G is torsion free then all units of $\mathcal{U}KG$ are trivial.

Very little progress has been made on this question. A solution to the special case when G is supersolvable will be most welcome. We state

Problem 11. If G is torsion free supersolvable then $\mathcal{U}\mathbb{Z}G = \pm G$.

Hughes and Pearson [1] computed $\mathcal{U}\mathbb{Z}S_3$, the unit group of the integral group ring of the symmetric group on 3 elements. Polcino Milies [1] did the same for the dihedral group of order 8.

Problem 12. Compute $\mathcal{U}\mathbb{Z}S_n$ and $\mathcal{U}KS_n$ for small values of n.

The next problem is due to Kaplansky [1].

Problem 13. $\gamma, \mu \in KG, \quad \gamma\mu = 1 \Rightarrow \mu\gamma = 1.$

This is known to be true if K has characteristic 0. See Jacobson [1], Losey [1], Peterson [1], Shepherdson [1] and Sehgal [7

We have next the famous isomorphism problem which is known to be true for Abelian groups, finite metabelian groups and some finite groups of small orders. See Passman [1], Cohn and Livingstone [1], Holovet [1],[2],[3], Saksanov [2], Jeya Kumar [1] and Miah [2].

Problem 14. Does $\mathbb{Z}G \simeq \mathbb{Z}H \Rightarrow G \simeq H$?

The same question may be asked for R or for fields. It is felt that relevant field for a p-group is $\mathbb{Z}/p\mathbb{Z}$. See Passman [2] and Dade [1].

Problem 15. Does $RG \simeq RH \Rightarrow G \simeq H$?

The following special cases are of interest.

Problem 16. Let G be a p-group. Does $(\mathbb{Z}/p\mathbb{Z})G \simeq (\mathbb{Z}/p\mathbb{Z})H \Rightarrow G \simeq H$?

Problem 17. Does G finite metabelian, $RG \simeq RH \Rightarrow G \simeq H$?

Saksanov [1] contains an incomplete proof of the above for the nilpotent class 2 case.

Problem 18. G metabelian (not necessary finite), $\mathbb{Z}G \simeq \mathbb{Z}H \Rightarrow G \simeq H$.

For the modular case, the isomorphism problem is known to have a positive solution if G is of M-class ≤ 2 (see (III.6.25)).

Problem 19. Let G be a p-group which is nilpotent of class 2. Then

$$(\mathbb{Z}/p\mathbb{Z})G \simeq (\mathbb{Z}/p\mathbb{Z})H \Rightarrow G \simeq H.$$

Problem 20. Suppose $\mathbb{Z}/p\mathbb{Z}G \simeq \mathbb{Z}/p\mathbb{Z}H$. Is there a normal subgroup corresponding between G and H?

Related to the isomorphism problem is Conjecture (III.7.5) whic we state below.

Problem 21. Let G be finite. Then every automorphism γ of $\mathbb{Z}G$ can be described as

$$\gamma(g) = \pm\alpha g^{\lambda}\alpha^{-1}, \quad g \in G$$

where λ is an automorphism of G and $\alpha \in \mathcal{U}QG$.

This is known to be true for nilpotent class two groups (III.7.4), S_n and certain classes of metabelian groups. See G. Peterson [1],[2], [3] and Brown [1]. The following special case should be decidable.

Problem 22. Last problem for metabelian groups.

A related problem is

Problem 23. Let G be finite. Every unit γ of $\mathbb{Z}G$ is of form $\gamma = \alpha g\alpha^{-1}$, $\alpha \in \mathcal{U}QG$.

See Hughes and Pearson [1].

Problem 24. Let G be a p-group. Does

$$(\mathbb{Z}/p\mathbb{Z})G \simeq (\mathbb{Z}/p\mathbb{Z})H \Rightarrow \mathbb{Z}_pG \simeq \mathbb{Z}_pH?$$

In this connection see (III.5.10). The next problem is known to have positive answer for finite groups (see (4.28)).

Problem 25. Does $\mathbb{Z}G \simeq \mathbb{Z}H$, G solvable (nilpotent) $\Rightarrow$ H solvable (nilpotent)?

The next problem is also a type of an isomorphism problem.

Problem 26. Let R_1, R_2 be unital rings and $\langle x\rangle$ an infinite cyclic group. Does

$$R_1\langle x\rangle \simeq R_2\langle x\rangle \Rightarrow R_1 \simeq R_2 \ ?$$

A positive answer to this in very special cases is given in Chapter IV. It is also proved that if R_1 and R_2 are commutative then $R_1\langle x\rangle \simeq R_2\langle x\rangle$ implies that R_1 and R_2 are subisomorphic. We propose the very special case.

Problem 27. Let R_1 be commutative Noetherian. Does

$$R_1\langle x\rangle \simeq R_2\langle x\rangle \Longrightarrow R_1 \simeq R_2?$$

Problem 28. What is the Jacobson radical of $R\langle x\rangle$?

See Amitsur [1], Herstein ([2], page 150) and Parmenter [2].

A problem related to the last one, namely, when does $R_1[x] \simeq R_2[x]$ imply $R_1 \simeq R_2$, has been studied by Coleman-Enochs [1], Abhyankar-Eakin [1] and others. Hochester [1] gave a counter-example to this. This example does not disprove Problem 25. Hence,

$$R_1[x] \simeq R_2[x] \not\Rightarrow R_1\langle x\rangle \simeq R_2\langle x\rangle.$$

We ask

Problem 29. Does $R_1\langle x\rangle \simeq R_2\langle x\rangle \Longrightarrow R_1[x] \simeq R_2[x]$?

Problem 30. Give a short proof of characteristic 2 case of Theorem (V.3.1).

We say an ideal I is nil of bounded exponent if there exists an integer n such that $x^n = 0$ for all $x \in I$. Let us consider $\Delta_K(G) = \Delta$. If Δ is nil of bounded exponent then certainly K has characteristic $p > 0$ and G is a p-group of bounded exponent Also, KG is Lie n-Engel. Hence G is nilpotent and contains a p-abelian normal subgroup A of finite p-power index. Conversely, it can be seen as in (V.6.7) that if $G \rhd A \rhd A'$ where G is a nilpotent p-group of bounded exponent, $(G:A) < \infty$, $|A'| < \infty$ and K has characteristic p then $\Delta_K(G)$ is nil of bounded exponent.

Problem 31. Characterize groups G such that $\Delta(G)$ is nil.

Connell [1] has partial results.

We know that if K has characteristic $p > 0$, G is nilpotent of class c and G' is a finite p-group then $(KG)^{[n]} = 0 = (KG)^{(m)}$ for some n and m.

Problem 32. Find a formula for n and m in terms of c and $|G'|$.

Problem 33. Characterize groups satisfying

$$(KG)^{(n)} = 0 \iff (KG)^{[n]} = 0.$$

Problem 34. Analogues of Problems 32 and 33 for $\delta^{[n]}(KG)$ and $\delta^{(n)}(KG)$.

The following problem is answered for integral group rings in (II.2.8).

Problem 35. Let A be a commutative unital ring. When is $(\mathcal{U}AG)^n \subset \zeta(\mathcal{U}AG)$?

In particular, one is interested in the case $A = K$.

Problem 36. Characterize groups such that $\mathcal{U}KG$ is n-Engel.

Problem 37. When is $\mathcal{U}KG$ an FC-group?

See Chapter VI for the rational and the integral cases.

We now mention a couple of questions not related directly to the material covered. We shall say that the intersection theorem holds for $\Delta = \Delta_A(G)$ if there exists a $\delta \in \Delta$ such that

$$\left(\bigcap_n \Delta^n\right)(1-\delta) = 0.$$

Sufficient conditions for the intersection theorem to hold for certain AG are given in Nouazé-Gabriel [1], Smith [1] and Parmenter-Sehgal [1]. In the last paper the following necessary and sufficient conditions are given.

(*) Let G be finitely generated with a nontrivial torsion element.

Let Π be the set of primes p such that G has an element of order p. Then the intersection theorem holds for $\Delta_{\mathbb{Z}}(G)$ if and only if for all $p \in \Pi$

$$\bigcap_n D_{n,\mathbb{Z}_p}(G)$$

is of finite order relatively prime to p.

(**) Let G be finitely generated. Then the intersection theorem holds for $\Delta_{\mathbb{Z}_p}(G)$ if and only if

$$\bigcap_n D_{n,\mathbb{Z}_p}(G)$$

is of finite order relatively prime to p.

(***) Let G be finite and I a commutative integral domain. Then the intersection theorem holds for IG if and only if

$$\left|\bigcap_n D_{n,I}(G)\right|$$

is invertible in I and every prime p dividing $|G|$ is invertible in I or satisfies $\bigcap_n p^n I = 0$.

<u>Problem 38</u>. Is there an extension of (***) to arbitrary groups?

<u>Problem 39</u>. Is there an extension of (*) to arbitrary groups? Can one remove the condition regarding existence of a torsion element?

Let us define $\Delta^{\omega} = \bigcap_n \Delta^n$ and by induction $\Delta^{\omega^m} = \left(\Delta^{\omega^{m-1}}\right)^{\omega}$ for natural numbers m. Write $\Delta^{\omega^{\omega}} = \bigcap_m \Delta^{\omega^m}$. Write Δ^{λ} for $\Delta^{\omega^{\omega}}$ and define by induction for natural numbers m

$$\Delta^{\lambda^m} = \left(\Delta^{\lambda^{m-1}}\right)^{\lambda} .$$

The following curious result is easily proved. See Parmenter-Sehgal [1].

<u>Theorem</u>. Let G be finite and let $\Delta = \Delta_{\mathbb{Z}}(G)$. Then $\Delta^{\lambda^m} = 0$ for a natural number m if and only if G is solvable.

<u>Problem 40</u>. Is there an extension of this result to more general groups?

The type of series of ideals defined above may be used to prove that if G is polycyclic by finite and $\mathscr{A}$ is an idempotent ideal contained in $\Delta_{\mathbb{Z}}(G)$ then $\mathscr{A} = 0$. It was conjectured by Akasaki [1] and proved by Roggenkamp [1] that for finite G, $\mathbb{Z}G$ contains a nontrivial idempotent ideal if and only if G is non-solvable. By using these two results Smith [2] has proved that if G is polycyclic by finite then $\mathbb{Z}G$ has nontrivial idempotent ideals if and only if G is non-solvable.

Problem 41. Characterize groups G such that $\mathbb{Z}G$ has no nontrivial idempotent ideals.

In this connection one should keep in mind (see Gruenberg [1]) that if G is an Abelian torsion divisible group then

$$\Delta_{\mathbb{Z}}^2(G) = \Delta_{\mathbb{Z}}^4(G) \neq 0.$$

We close with a problem of a different kind. In analogy with the L^1 group algebra we define

$$\ell(\mathbb{Z}_p, G) = \{\Sigma a_g g : a_g \in \mathbb{Z}_p, \lim a_g = 0\}.$$

See Parmenter-Sehgal [3] and Fresnel-Mathan [1]. It is well known that L^1 group algebra is semiprime. We ask

Problem 42. Is $\ell(\mathbb{Z}_p, G)$ semiprime?

BIBLIOGRAPHY

Abhyankar, S., Eakin, P. and Heinzer, W.

1. On the uniqueness of the coefficient ring in a polynomial ring, J. Algebra 23 (1972), 310-342.

Akasaki, T.

1. Idempotent ideals in integral group rings, J. Algebra 23 (1972), 343-346.
2. Idempotent ideals in integral group rings II, Arch. Math. 24 (1973), 126-128.

Amitsur, S.A.

1. Radicals of polynomial rings, Can. J. Math. 8 (1956) 355-361.

Artin, E.

1. Geometric algebra, Interscience, New York (1957).

Ayoub, R.G. and Ayoub, C.

1. On the group ring of a finite abelian group, Bull. Austral. Math. Soc. 1 (1969), 245-261.

Babakhanian, A.

1. Cohomological methods in group theory, Marcel Dekker, New York (1972).

Bass, H.

1. Finitistic homological dimension and a homological generalization of semi-primary rings, Trans. Amer. Math. Soc. 95 (1960), 466-488.
2. Euler characteristics and characters of discrete groups, Invent. Math. 35 (1976), 155-196.

Bateman, J.M.
1. On the solvability of unit groups of group algebras, Trans. Ame Math. Soc. 157 (1971), 73-86.

Bateman, J.M. and Coleman, D.B.
1. Group algebras with nilpotent unit groups, Proc. Amer. Math. Soc. 19 (1968), 448-449.

Bergman, G.M. and Dicks, W.
1. On universal derivations, J. Algebra 36 (1975), 193-211.

Berman, S.D.
1. On certain properties of the integral group rings, Dokl. Akad. Nauk, SSSR (N.S.) 91 (1953), 7-9.
2. On a necessary condition for the isomorphism of integral group rings, Dopovidi Akad. Nauk, USSR (1953), 313-316.
3. On the equation $x^m = 1$ in an integral group ring, Ukrainsk. Mat. Zh. 7 (1955), 253-261.
4. Group algebras of countable abelian p-groups, Soviet Math. Dokl 8 (1967), 871-873.

Berman, S.D. and Mollov, T. Zh.
1. On group rings of Abelian p-groups of any cardinality, Matem. Zametki 6 (1969), 381-392.

Bhattacharya, P.B. and Jain, S.K.
1. A note on the adjoint group of a ring, Arch. Math. 21 (1970), 366-368.

Bovdi, A.A.
1. The periodic normal divisors of the multiplicative group of a group ring, I. Sibirsk. Mat. Ž. 9 (1968), 495-498.
2. The periodic normal divisors of the multiplicative group of a group ring II, Sibirsk. Mat. Ž. 11 (1970), 492-511.
3. Group Rings, Uzgorod (1974).

Bovdi, A.A. and Khripta, I.I.
1. Normal subgroups of a multiplicative group of a ring, Mat. Sb. 87 (1972), 338-350; Math. USSR - Sb. 16 (1972), 349-362.
2. Finite dimensional group algebras having solvable unit groups, Trans. Science Conference Uzhgorod State University (1974), 227-233.

Bovdi, A.A. and Mihovski, S.V.
1. Idempotents in crossed products, Dokl. Akad. Nauk. SSSR 195; Soviet Math. Dokl. 11 (1970), 1439-1442.
2. Algebraic elements of crossed products, Colloquia Math. Soc. János Bolyai 6. Rings, modules and radicals; Keszthely, Hungary (1971), 103-116.

Brown, C.F.
1. Automorphisms of integral group rings, Thesis, Michigan State Univ. (1971).

Burns, R.G.
1. Central idempotents in group rings, Canad. Math. Bull. 13 (1970), 527-528.

Cliff, G.H. and Sehgal, S.K.
1. On the trace of an idempotent in a group ring, Proc. Amer. Math. Soc. 62 (1977), 11-14.

Cohen, D.E.
1. On abelian group algebras, Math. Z. 105 (1968), 267-269.

Cohn, J.A. and Livingstone, D.
1. On groups of order p^3, Canad. J. Math. 15 (1963), 622-624.
2. On the structure of group algebras, Can. J. Math. 17 (1965), 583-593.

Coleman, D.B.
1. Finite groups with isomorphic group algebras, Trans. Amer. Math. Soc. 105 (1962), 1-8.
2. On the modular group ring of a p-group, Proc. Amer. Math. Soc. 5 (1964), 511-514.
3. Idempotents in group rings, Proc. Amer. Math. Soc. 17 (1966), 962.
4. On group rings, Canad. J. Math. 22 (1970), 249-254.

Coleman, D. and Enochs, E.
1. Polynomial invariance of rings, Proc. Amer. Math. Soc. 25 (1970), 559-562.

Coleman, D.B. and Passman, D.S.
1. Units in modular group rings, Proc. Amer. Math. Soc. 25 (1970), 510-512.

Connell, I.G.
1. On the group ring, Canad. J. Math. 15 (1963), 650-685.

Curtis, C.W. and Reiner, I.
1. Representation theory of finite groups and associative algebras, Interscience, New York (1962).

Dade, E.C.
1. Deux groupes finis distincts ayant la même algebre de group sur toutcorps, Math. Z. 119 (1971), 345-348.

Dennis, R.K.
1. Units of group rings, J. Algebra 43 (1976), 655-664.
2. The structure of the unit group of group rings, Lecture notes in Pure and Applied Math. 26, Marcel Dekker, New York (1977).

Deskins, E.
1. Finite abelian groups with isomorphic group algebras, Duke Math. J. 23 (1956), 35-40.

Dieckman, E.M.
1. Isomorphism of group algebras of p-groups, Thesis, Washington Univ. (1967).

Dubois, P.F. and Sehgal, S.K.
1. Another proof of the invariance of Ulm's functions in commutati modular group rings, Math. J. Okayama Univ. 15 (1972), 137-139.

Eldridge, K.E.
1. On normal subgroups of modular group algebras. Preliminary report, Notices Amer. Math. Soc. 17 (1970), 764.

Fisher, J.L., Parmenter, M.M. and Sehgal, S.K.
1. Group rings with solvable n-Engel unit groups, Proc. Amer. Math Soc. 59 (1976), 195-200.

Formanek, E.
1. A short proof of a theorem of Jennings, Proc. Amer. Math. Soc. 16 (1970), 405-407.
2. Idempotents in Noetherian group rings, Can. J. Math. 25 (1973), 366-369.
3. The zero divisor question for supersolvable groups, Bull. Austral. Math. Soc. 9 (1973), 67-71.

Fresnel, J. and de Mathan, B.
1. Algebre du groupe infini et produit tensoriel de corps values, Séminaire de Théorie des Nombres, (1975-76), Univ. Bordeaux, Exp. No. 26, 34 pp., Talence (1976).

Fuchs, L.
1. Infinite abelian groups, volume 1, Academic Press, New York (1970).
2. Infinite abelian groups, volume 2, Academic Press, New York (1973).

Gruenberg, K.W.
1. Cohomological topics in group theory, Lecture notes in Math. Vol. 143, Springer-Verlag, New York (1970).

Hall, M.
1. The theory of groups, MacMillan, New York (1959).

Hall, P.
1. The Edmonton notes on nilpotent groups, Queen Mary College Math. Notes (1969).

Hasse, H.
1. Zahlentheorie, Akademie-Verlag, Berlin (1963).

Hattori, A.
1. Rank element of a projective module, Nagoya J. Math. 25 (1965) 113-120.

Herstein, I.N.
1. Topics in ring theory, Univ. Chicago Lecture Notes (1965).
2. Noncommutative rings, Carus Mathematical monographs, No. 15, Math. Assoc. Amer. (1968).

Higman, D.G.
1. On isomorphisms of orders, Mich. Math. J. 6 (1959), 255-258.

Higman, G.
1. The units of group rings, Proc. London Math. Soc. (2), 46 (1940), 231-248.

Hill, E.T.
1. Ideals in the modular group ring of a p-group, Proc. Amer. Math. Soc. 28 (1971), 389-390.

Hochester, M.
1. Non-uniqueness of coefficient rings in a polynomial ring, Proc. Amer. Math. Soc. 34 (1972), 81-82.

Holvoet, R.
1. Sur les Z_2-algèbres du groupe diédral d'ordre 8 et du groupe quaternionique, C.R. Acad. Sci. Paris, Sér. A, 262 (1966), 209-210.
2. Sur les algebres modulaires des groupes noncommutatifs d'ordre 12, C.R. Acad. Sci. Paris, Sér. A, 262 (1966), 690-692.
3. Sur l'isomorphie d'algebres des groupes, Bull. Soc. Math. Belg. 20 (1968), 264-282.
4. De groepalgebra van een eindige p-groep over een veld met karakteristiek p, Simon Stevin, 42 No. 4 (1969), 157-170.

Hughes, I. and Pearson, K.R.
1. The group of units of the integral group ring $\mathbb{Z}S_3$, Canad. Math. Bull. 15 (1972), 529-534.

Jackson, D.A.
1. The groups of units of the integral group rings of finite metabelian and finite nilpotent groups, Quart. J. Math. Oxford (2) 20 (1969), 319-331.

Jacobson, N.
1. Some remarks on one-sided inverses, Proc. Amer. Math. Soc. 1 (1950), 352-355.
2. Structure of rings, Amer. Math. Soc. Colloquium, vol. 36, Providence, R.I. (1964).

Janusz, G.J.
1. Algebraic number fields, Academic Press, New York (1973).

Jennings, S.A.
1. The structure of the group ring of a p-group over a modular field, Trans. Amer. Math. Soc. 50 (1941), 175-185.

2. On rings whose associated Lie rings are nilpotent, Bull. Amer. Math. Soc. 53 (1947), 593-597.
3. The group ring of a class of infinite nilpotent groups, Can. J. Math. 7 (1955), 169-187.
4. Radical rings with nilpotent associated groups, Trans. Roy. Soc Canad. 49 (1955), 31-38.

Jeyakumar, A.V.
1. Group algebras of dihedral and dicyclic groups over modular fields, J. London Math. Soc. 2 (1970), 297-304.

Kaplansky, I.
1. "Problems in the theory of rings" revisited, Amer. Math. Monthl 77 (1970), 445-454.

Khripta, I.I.
1. On the multiplicative group of a group ring, Thesis, Uzgorod (1971).
2. The nilpotence of the multiplicative group of a group ring, Mat Zametki 11 (1972), 191-200; Math. Notes 11 (1972), 119-124.

Kurosh, A.G.
1. The theory of groups, II, Chelsea Publ. Co., New York (1956).

Lambek, J.
1. Lectures on rings and modules, Blaisdell, Toronto (1966).

Lazard, M.
1. Sur les groupes nilpotents et les anneaux de Lie, Ann. Sci. Ecole Norm Sup. (3) 71 (1954), 101-190.

Losey, G.
1. Are one-sided inverses two-sided inverses in a matrix ring over a group ring? Canad. Math. Bull. 13 (1970), 475-479.
2. A remark on the units of finite order in the group ring of a finite group, Canad. Math. Bull. 17 (1974), 129-130.

May, W.
1. Commutative group algebras, Trans. Amer. Math. Soc. 136 (1969), 139-149.
2. Invariants for commutative group algebras, Ill. J. Math. 15 (1971), 525-531.
3. Group algebras over finitely generated rings, J. Algebra 39 (1976), 483-511.
4. Isomorphism of group algebras, J. Algebra 40 (1976), 10-18.

Miah, S.H.
1. Finite groups with isomorphic group algebras, Bull. Calcutta Math. Soc. 66 (1974), 111-112.
2. On the isomorphism of group algebras of groups of order 8q, J. London Math. Soc. 9 (1974/75), 549-556.

Montgomery, M.S.
1. Left and right inverses in group algebras, Bull. Amer. Math. Soc. 75 (1969), 539-540.

Moser, C.
1. Representation de -1 comme somme de carres dans un corps cyclotomique quelconque, J. Number Theory 5 (1973), 139-141.

Motose, K. and Ninomiya, Y.
1. On the solvability of unit groups of group rings, Math. J. Okayama Univ. 15 (1972), 209-214.

Motose, K. and Tominaga, H.
1. Group rings with nilpotent unit groups, Math. J. Okayama Univ. 14 (1969), 43-46.
2. Group rings with solvable unit groups, Math. J. Okayama Univ. 15 (1971), 37-40.

Nakajima, A. and Tominaga, H.
1. A note on group rings of p-groups, Math. J. Okayama Univ. 13 (1968), 107-109.

Newman, M.
1. Integral matrices, Academic Press, New York (1972).

Nouazé, Y. and Gabriel, P.
1. Indéaux primiers de l'algebre enveloppante d'une algebre de Lie nilpotente, J. Algebra 6 (1967), 77-99.

Obayashi, T.
1. Solvable groups with isomorphic group algebras, J. Math. Soc. Japan 18 (1966), 394-397.
2. Integral group rings of finite groups, Osaka J. Math. 7 (1970), 253-266.

Parmenter, M.M.
1. On a theorem of Bovdi, Can. J. Math. 23 (1971), 929-932.
2. Dimension theory, isomorphism and related problems in group rings, Thesis, University of Alberta (1972).
3. Isomorphic group rings, Canad. Math. Bull. 18 (1975), 567-576.
4. Coefficient rings of isomorphic group rings, to appear.

Parmenter, M.M., Passi, I.B.S. and Sehgal, S.K.
1. Polynomial ideals in group rings, Can. J. Math. 25 (1973), 1174-1182.

Parmenter, M.M. and Sehgal, S.K.
1. Idempotent elements and ideals in group rings and the intersection theorem, Arch. Math. 24 (1973), 586-600.
2. Uniqueness of the coefficient ring in some group rings, Canad. Math. Bull. 16 (1973), 551-555.
3. Non-archimedean group algebras, J. Number Theory 7 (1975), 376-384.

Pascaud, J.
1. Anneaux de groupes réduits, C.R. Acad. Sci., Paris, Sér. A, 277 (1973), 719-722.

Passi, I.B.S.
1. Dimension subgroups, J. Algebra 9 (1968), 152-182.

Passi, I.B.S., Passman, D.S. and Sehgal, S.K.
1. Lie solvable group rings, Canad. J. Math. 25 (1973), 748-757.

Passi, I.B.S. and Sehgal, S.K.
1. Isomorphism of modular group algebras, Math. Z. 129 (1972), 65-73.
2. Lie dimension subgroups, Commun. Algebra 3 (1975), 59-73.

Passman, D.S.
1. The group algebras of groups of order p^4 over a modular field Michigan Math. J. 12 (1965), 405-415.
2. Isomorphic groups and group rings, Pacific J. Math. 15 (1965), 561-583.
3. Infinite Group Rings, Marcel Dekker, New York (1971).
4. Group rings satisfying a polynomial identity III, Proc. Amer. Math. Soc. 31 (1972), 87-90.
5. Advances in group rings, Israel J. Math. 19 (1974), 67-107.
6. Algebraic structure of group rings, Interscience, New York (1977).
7. Observations on group rings, Comm. Algebra 5 (1977), 1119-1162.

Pearson, K.R.
1. On the units of a modular group ring, Bull. Austral. Math. Soc. 7 (1972), 169-182.
2. On the units of a modular group ring II, Bull. Austral. Math. Soc. 8 (1973), 435-442.

Perlis, P. and Walker, G.L.
1. Abelian group algebras of finite order, Trans. Amer. Math. Soc. 68 (1950), 420-426.

Peterson, G.
1. Automorphisms of the integral group ring of S_n, Proc. Amer. Math. Soc. 59 (1976), 14-18.
2. Automorphisms of integral group rings I, Arch. Math. (to appea
3. Automorphisms of integral group rings II, Ill. J. Math. 85 (1977), 836-844.

Peterson, R.D.
1. One-sided inverses in rings, Canad. J. Math. 27 (1975), 218-22

Polcino Milies, C.
1. The units of the integral group ring $\mathbb{Z}D_4$, Bol. Soc. Brasil. Mat. 4 (1972), 85-92.

2. Integral group rings with nilpotent unit groups, Can. J. Math. 28 (1976), 954-960.
3. p-adic group rings with nilpotent unit groups, J. Pure and Applied Algebra, to appear.

Quillen, D.
1. On the associated graded ring of a group ring, J. Algebra 10 (1968), 411-418.

Raggi Cárdenas, F.F.
1. Units in group rings I, An. Inst. Mat. Univ. Nac. Autonoma Mexico 7 (1967), 27-35.
2. Units in group rings II, An. Inst. Mat. Univ. Nac. Autonoma Mexico 8 (1968), 91-103.

Ribenboim, P.
1. Rings and Modules, Interscience, New York (1967).

Rips, E.
1. On the fourth integer dimension subgroup, Israel J. Math. 12 (1972), 342-346.

Robinson, D.J.S.
1. Finiteness conditions and generalized solvable groups, Ergebnisse der Math. vol. 62, 63, Springer Verlag, New York (1972).

Roggenkamp, K.W.
1. Integral group rings of solvable finite groups have no idempotent ideals, Arch. Math. 25 (1974), 125-128.

Rotman, J.
1. The theory of groups, Allyn and Bacon, Boston (1965).

Saksanov, A.I.
1. On group rings of finite groups I, Publ. Math. Debrecen 18 (1971), 187-209.
2. On an integral ring of characters of a finite group, Vestsi Akad. Nauk. USSR, Ser. Fiz.-Mat. (1966), 69-76.

Sandling, R.
1. The modular group rings of p-groups, Thesis, Univ. of Chicago (1969).
2. Fixed point free actions on p-groups, Arch. Math. 22 (1971), 231-236.
3. Dimension subgroups over arbitrary coefficient rings, J. Algebra 21 (1972), 250-265.
4. The dimension subgroup problem, J. Algebra 21 (1972), 216-231.
5. Note on the integral group ring problem, Math. Z. 124 (1972), 255-258.
6. Group rings of circle and unit groups, Math. Z. 140 (1974), 195-202.

Sehgal, S.K.
1. On the isomorphism of group algebras, Math. Z. 95 (1967), 71-7
2. On the isomorphism of integral group rings I, Canad. J. Math. 21 (1969), 410-413.
3. On the isomorphism of integral group rings II, Canad. J. Math. 21 (1969), 1182-1188.
4. Units in commutative integral group rings, Math. J. Okayama Univ. 14 (1970), 135-138.
5. On the isomorphism of p-adic group rings, J. Number Theory 2 (1970), 500-508.
6. On class sums in p-adic group rings, Can. J. Math. 23 (1971), 541-543.
7. Lie properties in modular group algebras; Orders, Group Rings and Related Topics (Proc. Conf., Ohio State Univ., Columbus, Ohio), Lecture Notes in Math., vol. 353, Springer-Verlag, New York (1973).
8. Certain algebraic elements in group rings, Arch. Math. 26 (197 139-143.
9. Nilpotent elements in group rings, Manuscripta Math. 15 (1975) 65-80.

Sehgal, S.K. and Zassenhaus, H.J.
1. Group rings whose units form an FC group, Math. Z. 153 (1977), 29-35.
2. Integral group rings with nilpotent unit groups, Commun. Algebra 5 (1977), 101-111.
3. Group rings without nontrivial idempotents, Arch. Math. 28 (1977), 378-380.

Shepherdson, J.C.
1. Inverses and zero divisors in matrix rings, Proc. London Math. Soc. 3 (1951), 71-85.

Sjogren, J.A.
1. Thesis, University of California, Berkeley (1975).

Smith, P.F.
1. On the intersection theorem, Proc. London Math. Soc. 21 (1970) 385-398.
2. A note on idempotent ideals in group rings, Arch. Math. 27 (1976), 22-27.

Tahara, K.
1. The fourth dimension subgroups and polynomial maps, J. Algebra 45 (1977), 102-131.
2. The fourth dimension subgroups and polynomial maps, II, Nagoya Math. J. 69 (1978), 1-7.

Ward, H.N.
1. Some results on the group algebra of a group over a prime field, Harvard seminar on finite groups, 1960-61.

Whitcomb, A.
1. The group ring problem. Ph.D. Thesis, University of Chicago (1968).

Williamson, A.
1. On the conjugacy classes in a group ring, Canad. Math. Bull. (to appear).

Zalesskii, A.E.
1. On a problem of Kaplansky, Dokl. Akad. Nauk. SSSR 203 (1972), 749-751 (Russian); Soviet Math. Dokl. 13 (1972), 449-452.

Zalesskii, A.E. and Mikhalev, A.V.
1. Group rings, Itogi Nuaki i Tekhniki (Sobremennye Problemy Matematiki) 2 (1973), 5-118; J. Soviet Math. 4 (1975), 1-78.

Zariski, O. and Samuel, P.
1. Commutative algebra, I, Van Nostrand, Princeton (1958).

Zassenhaus, H.
1. Ein Verfahren jeder endlichen p-Gruppe einen Lie-Ring mit der Charakteristik p zuzuordnen, Abh. Math. Sem. Univ. Hamburg 13 (1940), 200-207.
2. On the torsion units of finite group rings, Studies in Mathematics (in honor of A. Almeida Costa), Instituo de Alta Cultura, Lisbon (1974), 119-126.
3. When is the unit group of a Dedekind order solvable? Commun. Algebra, to appear.

Zhmud, E.M. and Kurennoi, G.C.
1. On finite groups of units of an integral group ring, Vesnik Har'kov. Gos. Univ. 26 (1967), 20-26.

Ziegenbalg, J.
1. The lattices of the finite normal subgroups in group bases of isomorphic integral group rings are isomorphic, Arch. Math. 26 (1975), 231-233.
2. Isomorphe Gruppenringe lokal endlicher Gruppen, J. reine angew. Math. (to appear).

INDEX OF SYMBOLS

INDEX